Everyday Mathematics®

The University of Chicago School Mathematics Project

Resources for the Kindergarten Classroom

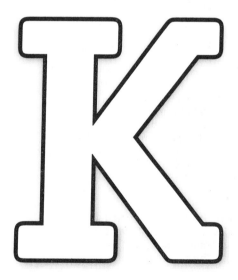

Kindergarten

McGraw Hill Education

The McGraw·Hill Companies

The University of Chicago School Mathematics Project (UCSMP)

Max Bell, Director, UCSMP Elementary Materials Component, Director, *Everyday Mathematics* First Edition; James McBride, Director, *Everyday Mathematics* Second Edition; Andy Isaacs, Director, *Everyday Mathematics* Third Edition; Amy Dillard, Associate Director, *Everyday Mathematics* Third Edition; Rachel Malpass McCall, Associate Director, *Everyday Mathematics* Common Core State Standards Edition

Authors
Ann E. Audrain, Margaret Krulee, Deborah Arron Leslie, Barbara Smart

Third Edition Early Childhood Team Leaders
David W. Beer, Deborah Arron Leslie

Technical Art
Diana Barrie

UCSMP Editorial
Patrick Carroll, Tiffany Nicole Slade

Contributors
Dorothy Freedman, Juanita V. Copley, Michelle Patt

Permissions
The Hokey Pokey, ©1950 Sony/ATV Songs LLC.
All rights administered by Sony/ATV Music Publishing,
8 Music Square West, Nashville, TN 37203. All rights
reserved. Used by permission.

Photo Credits
Front cover (l)Willard R. Culver/National Geographic/Getty Images (r)Brand X Pictures/PunchStock, (bkgd)Jules Frazier/Photodisc/Getty Images; **Back Cover** Willard R. Culver/National Geographic/Getty Images; **67** The McGraw-Hill Companies; **87** Image Source/Alamy; **88 91** The McGraw-Hill Companies.

everyday**math**.com

 **Education**

STEM McGraw-Hill is committed to providing instructional materials in Science, Technology, Engineering, and Mathematics (STEM) that give all students a solid foundation, one that prepares them for college and careers in the 21st century.

Send all inquiries to:
McGraw-Hill Education
STEM Learning Solutions Center
P.O. Box 812960
Chicago, IL 60681

ISBN: 978-0-07-657584-8
MHID: 0-07-657584-5

Printed in the United States of America.

1 2 3 4 5 6 7 8 9 QDB 17 16 15 14 13 12 11

Contents

Introduction

Everyday Mathematics Resources for the Kindergarten Classroom is a collection of materials and resources that supplement the activities in the *Teacher's Guide to Activities*. This book suggests ways to reinforce, enrich, and extend children's mathematics experiences, with an emphasis on integrating mathematics into all aspects of the classroom. This book reflects the *Everyday Mathematics* philosophy: young children should have many opportunities to explore mathematical concepts through play, informal interactions with adults and other children, and through exposure to a variety of materials and manipulatives in a range of contexts. Choose any of these optional activities that appeal to you and your class. The book includes the following sections:

Mathematics All Around (pp. 1–4)

The descriptions in this section are intended to help teachers recognize and value the mathematics that is embedded in children's work and play. This section includes examples of the many ways that children naturally use mathematics in their outdoor play, their artwork, their science explorations, and their activities in the dramatic play and block areas. These descriptions may help you articulate to parents, administrators, or others some of the ways that children learn and apply mathematics through their spontaneous work and play in the classroom and at home.

Theme Activities (pp. 5–41)

This section highlights mathematics play and activities that can be incorporated into eight commonly used thematic units. The theme activities in this book focus on mathematics activities only; they do not suggest activities in other curricular areas that could also be part of these themes. There are many other good resources available that suggest thematic activities in other curricular areas.

Songs and Poems (pp. 42–48)

This section includes lyrics and text for a variety of mathematics-based songs, poems, chants, and fingerplays that can be used during music time, group times, or transition times. Some of the selections have been suggested as part of the Teaching Options in the *Teacher's Guide to Activities.*

Literature Lists (pp. 49–59)

These lists include a variety of children's books that can be used to reinforce and enrich children's understanding of mathematical concepts. The lists are organized by topic, although not all of the books are strictly mathematical. (Many are just good stories that can be used to illustrate concepts or promote discussion about a particular mathematics topic.) Many of the books on the lists are referenced in activities in the *Teacher's Guide to Activities*. A list of resource books for teachers is also included.

Software List (pp. 60–61)

This section highlights age-appropriate computer software that reinforces many of the topics children will learn about through *Kindergarten Everyday Mathematics*.

Games for the Classroom (pp. 62–65)

This section highlights some board games, card games, and other games that children can play in the classroom to reinforce mathematics concepts and skills. (*Kindergarten Everyday Mathematics* also includes numerous games that can be made in the classroom.)

Ideas for Newsletters (pp. 66–72)

This section includes short, parent-friendly descriptions about the mathematics you will do in class. You can incorporate these suggestions into your newsletters or other home/school communications.

Family Letter Masters (pp. 73–78)

This section includes blackline masters for four letters that can be sent home to families to explain *Kindergarten Everyday Mathematics*. The first letter introduces the program and should be sent home near the beginning of the year. The second letter explains the Ongoing Daily Routines. It can be sent home about a month into the school year. The third letter can be sent home at about midyear, and the fourth letter (which includes suggestions for summer mathematics activities) should be sent home near the end of the school year.

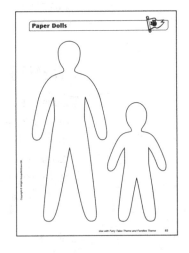

Theme Masters (pp. 79–85)

These blackline masters can be used in conjunction with some of the theme activities described in this book.

Mathematics All Around

Young children spontaneously explore mathematics all the time. They count and sort, notice and describe shapes and patterns, estimate and compare sizes, and wonder about the numbers they see all around them. These mathematical observations, concepts, and skills are important tools that children use as they explore and make sense of their world.

Children learn a great deal as they pursue their own interests and initiate activities. By recognizing what children are already doing with mathematics, you will be better equipped to respond to their spontaneous mathematical activities in ways that foster continued exploration and learning. Sometimes the best response involves simply watching what a child is doing, without intervening at all. Sometimes it means engaging the child in conversation or asking a thought-provoking question. Sometimes it means drawing other children in or, if the child is receptive, building on or extending a spontaneous activity. In each case, take care not to interrupt children's spontaneous play with an abundance of questions, instructions, or teacher-directed tasks.

The following lists describe typical activities young children do in the early childhood classroom, around the school, and outdoors. They are intended to help you recognize opportunities to observe and encourage children's natural explorations of mathematics throughout the school day. These lists may also help you consider ways to set up or modify your classroom environment to promote natural mathematics explorations.

Mathematics Outdoors

Mathematics opportunities are abundant in outside play. They emerge naturally from games and active play, as well as from children's interactions with nature. Listen carefully during recess and you will hear children exploring mathematical concepts spontaneously: *How many jumps? How high can you make it bounce? Throw it farther! Watch how fast I can run! Let's dig a really deep hole. Which flower is taller?* The following are just a few more examples of children's spontaneous mathematics explorations outdoors.

▷ Children **count objects and actions:** rungs on the monkey bars, the number of children on the climber, or the number of pushes on the swing.
▷ Children **notice and describe shapes** all around them. (*The hopscotch spaces are squares. The ball is round* (a sphere). *The traffic signs have different shapes.*) Children may create shapes with chalk or with their bodies.
▷ Children **investigate and compare linear measurements** such as height, length, or distance. (*Which slide is longer? Why is my shadow so tall? I'm swinging higher than the trees!*)

- Children **think and talk about time and speed.** (*How many minutes until we go inside? Let's race the balls down the slide!*)
- Children **use position vocabulary** such as: *on* the bridge, *in* the tunnel, *back and forth* on the swing, *high* in the sky, and *under* the leaf.
- Children **notice and describe patterns** in their surroundings: the up and down of a bouncing ball, the color patterns of flowers, the arrangement of bricks on a path, or the patterns in bird calls.
- Child naturalists and collectors **sort objects by attributes**, including their rock, leaf, or other collections.

Mathematics and Art

Shapes, patterns, size, scale, and symmetry are just a few of the important concepts common to both mathematics and art. For children who are drawn to art or craft activities, the list below may provide ideas for embedding mathematical learning into their explorations in the Art Center. It may also help you find ways to bring reluctant children to the Art Center by capitalizing on their 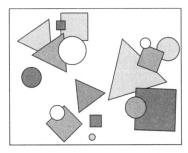 interest in counting, patterns, measurement, or other mathematical concepts. Both the art and the mathematics will be more meaningful to children if they can pursue ideas that emerge from their own creativity and exploration rather than projects that require all children to produce identical products. The following are just a few examples of children's spontaneous mathematics explorations through art:

- Children **count and think about quantities** when creating artwork. (*How many beads will I need? I used 10 toothpicks and 10 marshmallows.*)
- Children **compare quantities** in artwork. (*I have more buttons than beads.*)
- They **incorporate and describe shapes.** (*The house is made up of a triangle and a square. The overlapping circles make an interesting design.*)
- Children **explore and compare measurements.** (*How much yarn do I need? Look how long my chain is!*)
- Children **develop position and relationship concepts** as they compose and describe their artwork. (*I am standing between my mom and dad. I drew my sun at the top of the paper. The flowers are next to the house.*)
- Children **create patterns** in their artwork, especially if multiple colors, textures, or materials are available for them to use.
- Children **sort and classify materials or objects** when doing artwork. They might make a picture using only the color green, or using only shiny buttons for a collage.

Mathematics and Science

Science and mathematics are closely linked; both focus on understanding and describing objects and events in the environment. Science provides a meaningful context in which children can apply their developing mathematics skills: sorting a collection of rocks to learn more about their attributes; noticing patterns in the natural world; figuring out how long it takes a ball to roll down a ramp; and measuring how long the ramp is. Young children are natural scientists—curious and observant, as they take in and make sense of their world. Many of children's scientific observations have a mathematical component, such as attention to size, shape, or quantity. Here are just a few more examples of how children use mathematics in their scientific explorations:

▷ Children **count and compare** as they make observations. (*This spider has 8 legs; that ant only has 6. My flower has 6 petals and 4 leaves.*)

▷ Children **notice and describe shapes and symmetry** when they look at things closely: the circular center of a flower, the triangular beak of a bird, the symmetrical wings of a butterfly.

▷ Children **compare, describe, and measure height, length, and weight**. (*The big shell is heavier than the small shell. My plant is 5 inches tall!*)

▷ Children **think about time concepts** as they observe and describe events and changes. (*The tree lost its leaves during the winter. My plant grew 2 inches in 2 days. Count how long it takes for the rock to drop to the ground.*)

▷ They **estimate amounts and measurements** as they work with materials. (*I think I will need 2 cups to hold all the sand. I'll need a longer pole to hold up my plant.*)

▷ Children **notice and describe patterns and symmetry** in nature and in physical phenomena such as striped animal markings, the back and forth of a moving swing, or a vibrating rubber band.

▷ Children **sort natural and human-made objects** according to different attributes. Sorting encourages children to look closely and notice details.

Mathematics and Dramatic Play

There are many ways in which mathematics is integrated into dramatic play. One-to-one correspondence, number sense, sorting, measurement, size, and scale are pervasive as children set the table, play store or restaurant, dress up, pretend to cook, and engage in other imaginary play. Rotate and add materials that will help children extend the mathematics that might emerge from their play. Here are just a few more examples of children's spontaneous mathematics explorations through dramatic play:

▷ Children **identify and use numbers** when playing with telephones, keyboards, cash registers, cereal boxes, and so on. They naturally use ordinal numbers to organize their play. (*I get to do the cash register first. What should we add to the pot next? You can be last.*)

▷ Children **count and compare numbers of objects** during their play. (*How many more cups do we need? I only have 3 shoes!*)

▷ Children **use one-to-one correspondence** as they set the table or serve food.
▷ Children **use money** when playing in a pretend store or restaurant or when making and selling tickets for a carnival or show.
▷ Children **explore measurement concepts and vocabulary** as they pretend to cook food. (*I added 7 cups of salt! Bring me 2 pounds of potatoes.*) Including measuring tools in the Dramatic Play Center will promote this exploration.
▷ Children **describe and compare sizes** as they pretend to serve food or dress up. (*I need a bigger piece of pizza. My dress is longer than yours.*)
▷ Children **incorporate time concepts and language** into their play. (*Time for bed! It's 7:00. Let's pretend it's next year, and we're in first grade.*)
▷ Children **use position vocabulary** to describe the relationships of objects and children. (*Sit next to me! Put the soup on the table. Go under the bridge.*)
▷ Children **sort and categorize materials** as they play and during clean-up. (*Put all the girls' dress-up clothes in one box. All the vegetables go together.*)

Mathematics in the Block Center

Mathematics is an integral part of block play, and children's mathematical learning in the block area is deep and lasting because it develops over time through repeated, concrete experiences. As children build, they have first-hand experiences with 3-dimensional shapes and important spatial relationship concepts. They experiment with height, weight, and balance as they make stable structures and explore shape, symmetry, and patterns as they design buildings. Children get immediate feedback in their efforts to problem solve: they can see that the garage is not big enough for the cars, or the doll can't fit through the doorway of the house. Building encourages children to think flexibly and mathematically to solve problems. Here are a few ways children become actively engaged in mathematics during block play:

▷ Children **count and compare quantities** as they build. (*This side has 6 blocks and this one has 4. I have to find 2 more blocks before I can add the roof.*)
▷ Children **use numerals** to label buildings with addresses or floor numbers.
▷ Children **become familiar with 3-dimensional shapes** as they build with blocks and talk about the blocks they are using. They also **notice and describe 2-dimensional shapes** that make up the faces of the blocks. As they manipulate blocks, children come to understand that rotating or repositioning objects does not change their shape—an important concept in geometry.
▷ Children **explore relationships among shapes and sizes** as they find different ways to combine blocks to produce desired shapes and sizes.
▷ Children **incorporate symmetry and patterns** into their buildings.
▷ Children **measure and compare constructions** using both standard and nonstandard measurement tools and units. (*My skyscraper is almost as tall as you! This garage comes up to my knees. Our castle is as long as 4 rulers!*)
▷ Children **discover the importance of weight and balance** in building. (They learn that their structures will fall if they are top-heavy or unbalanced!)
▷ Children **use position words** to describe their actions or to collaborate with other builders. (*Put that one on top. I'm putting the cylinder next to the door.*)
▷ Children **sort according to size and shape** as they put blocks away.

Theme Activities

Introduction to Theme Activities

Many Kindergarten teachers use thematic units as part of their teaching. Although teachers use different themes (and no two teachers approach the same theme in exactly the same way!), this section suggests some mathematics activities that can be done in conjunction with the following eight common Kindergarten themes:

▷ A Working World   ▷ Dinosaurs   ▷ Growing Things
▷ All About Me   ▷ Fairy Tales   ▷ Seasons
▷ Animals All Around   ▷ Families

If you use any of these themes in your classroom, consider the activities in this section as suggestions that you can modify to suit your class, teaching style, and approach to each theme. All of these activities are optional in *Kindergarten Everyday Mathematics*. If thematic units are not part of your instructional approach, or if you do not use a particular theme, you can still use individual theme activities to reinforce mathematical concepts throughout the year.

Blackline masters to support specific theme activities are included at the end of this book beginning on page 79. The books that are referenced in the theme activities are included on the literature lists beginning on page 49.

Mathematics and Other Themes

This book includes only a few of the many different themes that Kindergarten teachers frequently use. Similarly, although the theme activities included here are aimed at emphasizing mathematics, the list of mathematics activities related to each theme is not exhaustive. The ideas in this section may give you ideas for new themes or activities to try in your classroom. As you look for opportunities to use mathematics in these and other themes, consider the following:

▷ Gameboards can be created to accompany any theme or area of study. Games that involve using dice or spinners to move pieces along a path reinforce counting and one-to-one correspondence.

▷ Recording, counting, graphing, and comparing data can be part of virtually any theme, using questions or topics that relate to the theme.

In all cases, consider how you might coordinate mathematics activities with theme activities related to other curricular areas. (The best activities usually span several areas!) Throughout the year, be prepared to adjust your plan for thematic units to address emerging interests.

A Working World

Children enjoy pretending to do "grown-up" jobs. They see family members go to work and notice people who work in their schools and neighborhoods. Family members and other adults are an excellent resource for learning about different occupations. The activities described in this section incorporate various mathematics concepts and skills as children learn about and role play different kinds of jobs. Many of the activities can be incorporated into a Dramatic Play Center that explores jobs and workplaces.

Mathematics at the Post Office

Strands Number and Numeration; Measurement and Reference Frames; Patterns, Functions, and Algebra

Focus Explore weighing, sorting, and numbers in children's surroundings.

▷ **Weighing the mail** Children can use a pan balance and uniform weights (such as pennies, counters, or paper clips) to weigh different-sized envelopes and packages and compare weights. Discuss why mail is weighed (to find out how many stamps to use). Invite children to create paper stamps (or use rubber stamps and ink pads) to put on pretend letters and packages of various sizes. Discuss other jobs that require workers to weigh things. What tools do they use for weighing?

▷ **Sorting the mail** Provide—or have children create—letters and packages of different sizes, shapes, or colors. You might also add pictures to indicate the destination of each package. Invite children to sort the mail in different ways. Explain that mail is sorted at the post office according to where it is going. (You might use the addresses on some real letters to explain that the addresses tell where the mail should be delivered.)

▷ **Learning about addresses** Help children learn their addresses. Encourage them to notice other addresses around their home or neighborhood. Do they notice anything about the numbers? Talk about what the numbers in an address mean. (Addresses are *codes*. For more information about codes, see Project 1, Numbers in Our World, on page 74 of the *Teacher's Guide to Activities*.) You might add numbers to children's lockers or cubbies and invite them to use these numbers to write addresses on letters to each other. Children can deliver their own mail, or you might designate a classroom mailbox and a letter carrier for each day!

NOTE *The Jolly Postman* by Allan Ahlberg (Little, Brown, 2001) and *Dear Peter Rabbit* by Alma Flor Ada (Aladdin, 1997) are books that might inspire children to write letters to one another. Encourage them to notice the addresses, postage, and other numbers on the mail in these books.

Mathematics and Building

Strands Measurement and Reference Frames; Geometry

Focus Explore shapes, spatial relationships, and measurement while building.

▷ **Planning to build** Talk with children about the different tasks that are required to design and build a house or other structure. Before they build with blocks or other materials, you might encourage them to think about what they want their buildings to look like, which block shapes and sizes they will use, how they will put them together, and so on. Explain that architects draw blueprints to show their plans for buildings, and invite children to draw blueprints for their buildings. (Display blueprints next to completed buildings.)

▷ **Building** As children build, talk with them about the sizes and shapes of the blocks that they use. If they run out of one type of block, encourage them to put together other blocks to make the size and shape they need. Invite them to compare blocks, orient them in different ways, and use position words to describe what they are doing (*Do you want to put that block between these two?*). Talk with them about how their work compares to the real work of architects and builders. (Professionals must also consider the size, shape, position, and orientation of building materials. They also may need to make adjustments and solve problems as they build.)

▷ **Measuring buildings** After children construct a building, invite them to use measurement tools to measure their structures. Add rulers, tape measures, and yardsticks to the materials in the block area, and encourage children to incorporate them into their play. Ask them if they know of any other tools that architects or construction workers use. Children can record and compare the sizes of different buildings. They might discover that you can measure more than one dimension (height and width, for example).

Mathematics at the Doctor's Office

Strands Number and Numeration; Measurement and Reference Frames

Focus Practice counting, measuring, weighing, and recording numbers in a pretend doctor's office.

Invite children to measure the height of a doll, a stuffed animal, or a classmate "patient" using a height chart, ruler, or yardstick. If possible, provide a scale to measure weight. (A doll or stuffed animal will probably not be heavy enough to register weight on a bathroom scale.) Provide a clipboard and paper for children to record height and weight measurements. Children can also record other numbers on the patient's "chart," such as the patient's age or number of loose teeth. You might show children how to count their pulse rate by holding their hand over their heart or putting a finger on their wrist or the pulse point on their neck. (The pulse is easier to detect after doing jumping jacks or some other exercise.) If possible, invite the school nurse to talk with the class and explain how numbers, measurements, or other mathematics skills are used in his or her work.

NOTE Children may check their pulse as part of Project 2, Mathematics and Our Bodies, on page 127 of the *Teacher's Guide to Activities.*

Setting Up Shop

Strands Number and Numeration; Measurement and Reference Frames

Focus Practice counting, writing numbers, and using money to play store.

Talk with children about items that people buy and sell. Have the class brainstorm a list of the mathematics skills that people who work in stores must use every day. You can create a store in the Dramatic Play Center and invite children to think of something that the class can sell. Set up the store by putting out items and having children sort and arrange them, as well as make price tags and "For Sale" signs. Encourage children to discuss how many coins are needed to buy and sell items. (Keep the prices low!) Children can take turns being the cashier using real or pretend money. Children can also share the cashier's job with a partner.

You might want to have children bring in old toys from home to resell for a penny, create items to sell, or open a grocery store by adding food items that can be sold to other children in the school.

NOTE Children begin working with coins in Section 6 of the *Teacher's Guide to Activities*.

Bakery Mathematics

Strands Number and Numeration; Measurement and Reference Frames; Patterns, Functions, and Algebra

Focus Apply a variety of mathematics skills in the context of cooking and setting up a bakery.

Cooking involves lots of mathematics! Children can count and measure ingredients, set the oven temperature, set a timer or watch the clock to monitor cooking time, and figure out a fair way to distribute the finished product. You might use the pictures of measuring tools from *Math Masters*, pages 77 and 78, to make a pictorial recipe that children can "read" and follow.

In a pretend bakery, children use mathematics skills as they sort and arrange items on the shelves, set prices and make price tags, and exchange pretend money to buy and sell baked goods. Encourage children to think about (and try to find out) the number of eggs, bags of flour, and other ingredients that a real bakery must use each day!

NOTE See Project 4, Class Celebration, on page 225 of the *Teacher's Guide to Activities* for recipes you might make in class.

Mathematics in a Restaurant

Strands Number and Numeration; Measurement and Reference Frames; Geometry; Patterns, Functions, and Algebra

Focus Explore a variety of mathematics skills in setting up and working in a pretend restaurant.

Many of the mathematics activities involved in learning about stores and bakeries also apply to restaurants. In addition, if children set up a classroom restaurant, they might write numbers and money amounts as they create menus, practice one-to-one correspondence as they set tables, and explore shapes and patterns as they make tablecloths, placements, or decorations.

NOTE Many of the ideas in Project 4 can be modified to set up a class restaurant instead of a class celebration. Help children recognize that there are many different jobs involved in running a restaurant (and every other establishment).

Sorting Work Tools

Strand Patterns, Functions, and Algebra

Focus Sort a variety of tools in different ways.

Discuss with children the different tools or objects that people use in their jobs. Talk about the various types of things that can be classified as tools, such as a hammer, shovel, paper clip, stapler, computer, stethoscope, scissors, measuring cup, or anything else that helps people do their jobs. Collect a wide assortment of tools for children to sort in various ways. Have children explain how they sorted. Children can supplement your tool collection by bringing tools from home, drawing pictures of tools, or cutting pictures of tools out of magazines. You might want to place the tools in the Dramatic Play Center for children to use in their play.

"When I Grow Up" Graph

Strand Data and Chance

Focus Create and analyze a class bar graph about children's future career choices.

Have children draw a picture of a job they want to do when they get older. Children can draw on a stick-on note or small index card. You can use the notes or cards to create a bar graph that shows how many children are interested in doing different jobs. (In the course of the discussion, be sure to convey that children will have plenty of time to think about jobs and to change their minds many times as they grow older. For this activity they should just think about something that sounds interesting right now.) Use the graph as the basis for counting, comparing numbers, and creating number stories.

NOTE Children make bar graphs in Activities 1-8 (Birthday Graphs), 3-14 (Favorite Colors Graph), and 5-13 (Pet Bar Graph), as well as several others. You might want to refer to those activities for more specific suggestions about creating class bar graphs.

Using Mathematics at Work

Strands All strands

Focus Think about the many ways that mathematics is used in different jobs.

At the end of a theme or unit about the working world, you might ask children to think of all the ways that numbers are used in the jobs they learned about. List their ideas. You can repeat for other mathematical topics, such as measurement, shapes, or problem-solving. Ask children how they or their family members use mathematics at home, too!

All About Me

An "All About Me" theme offers many meaningful opportunities for mathematical practice and applications, especially in the areas of measurement, number and numeration, and sorting and graphing attributes.

Making Handprints

> **Strands** Number and Numeration; Measurement and Reference Frames; Patterns, Functions, and Algebra
>
> **Focus** Explore patterns, counting, and measurement through an art activity.

Children can dip their palms in paint to make handprints on long strips of paper. Provide a variety of colors for children to make handprint patterns. Ask: *What color comes next in your pattern? Can you make a different pattern?* Encourage mathematical thinking and discussion by asking: *How many handprints fit on the paper? How many fingers are on your paper? How does the size of your hand compare to the size of my hand? Which of us could fit more handprints on the paper?*

Measuring Height

> **Strand** Measurement and Reference Frames
>
> **Focus** Use nonstandard measurement to measure and compare heights.

Children can measure their height with blocks. One child lies down while the other forms a line of matching unit blocks next to him or her. As children measure, ask: *How many blocks have you used so far? How many more blocks do you think you will need?* Children count the number of blocks in the line to determine "about how many blocks tall" the measured child is. (Reinforce that the line of blocks may not **exactly** match the length of the child.) Children can record their measurements by writing their names and the date and drawing and/or writing the number and type (shape and size) of blocks used. They might repeat the measurement at different times and record these measurements and dates to document their growth over time.

Children can also use other materials to measure their height. They can compare the results of measuring with blocks and measuring with other items.

NOTE This is similar to Activity 3-7, Measurement with Objects, in the *Teacher's Guide to Activities*.

Body Tracings

> **Strands** Measurement and Reference Frames; Geometry; Patterns, Functions, and Algebra
>
> **Focus** Use body tracings to think and talk about symmetry, measurement, shapes, and patterns.

With partners, children can take turns tracing around each other's bodies, either with a marker on bulletin-board paper or with chalk on a sidewalk or blacktop. Have children fill in their body outlines with facial features and clothes. Use the body outlines as the source for mathematical discussions and activities, such as:

▷ Discussion of symmetry in their bodies (whole body, face, nose, and so on),
▷ Measurement of arm length, leg length, or overall height,
▷ Comparisons of body parts to geometric shapes, such as triangular noses and oval eyes,
▷ Addition of patterned clothing on the body outlines.

Measuring and Comparing Shadows

> **Strand** Measurement and Reference Frames
>
> **Focus** Explore measurement through shadow play and experimentation.

Use the poem "My Shadow," by Robert Louis Stevenson, to initiate this activity. Here is the first stanza:

> *I have a little shadow that goes in and out with me*
> *And what can be the use of him is more than I can see*
> *He is very, very like me from the heels up to the head;*
> *And I see him jump before me, when I jump into my bed.*

In a darkened room, pairs or small groups of children can shine flashlights on themselves and each other to create shadows. Encourage children to discover ways to change the size of their shadows. Ask: *What happens if you move the flashlight close to your partner? Whose shadow is bigger?* Mark the outline of the shadows with tape and invite children to measure them. Depending on the measurement activities you've done so far, children might measure with their feet (or other body part), with nonstandard uniform units such as connecting cubes or blocks, or with standard measuring tools. Children can also work with shadows outdoors on a sunny day. They might compare the size of their shadows with the size of other objects' shadows.

NOTE This is similar to shadow activities in Project 8, Math Outdoors, pages 417 and 418 in the *Teacher's Guide to Activities*.

Matching Birth Weights

Strand Measurement and Reference Frames

Focus Conduct concrete investigations to develop a sense of weight.

Children can place an empty bag on a bathroom scale, then load the bag with objects until it weighs as much as they weighed when they were born. Let children carry the bag around to get a feel for the weight of a newborn baby. Children might do a similar activity at home to match their present weight. They can draw or otherwise record the collection of objects that matches their birth and/or present weight.

Graphing Features, Favorites, and Other Data

Strand Data and Chance

Focus Practice data collection and graphing to explore similarities and differences.

Help the class make various bar graphs that depict the variety in children's features, such as eye color or hair color. Enlist children's help in deciding how to set up the graphs and how to collect the data, and in choosing labels and a title for the graphs. Children can also graph the number of letters in their names, number of siblings, or various "favorites," such as favorite color or ice cream flavor.

NOTE Children make bar graphs in Activities 1-8 (Birthday Graphs), 3-14 (Favorite Colors Graph), and 5-13 (Pet Bar Graph), as well as several others. You might want to refer to those activities for more specific suggestions about creating class bar graphs.

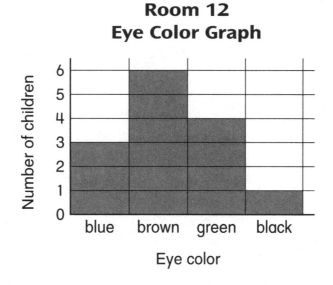

Making "My Day" Books

Strand Measurement and Reference Frames

Focus Explore time and sequence in the context of children's daily schedules.

Children can write or dictate, then illustrate, brief stories about what they do during different parts of the day. Have children put their pictures in the order the events occur, and help them add time or time-of-day language to the top of each page (for example, morning, afternoon, evening; before school, during school, after school, 9:00, 12:00, 3:00). Assemble each child's pages into a book, and have the child choose the title.

As a variation, the class can collaborate on a book about a typical school day. In some classrooms, children make a book such as this at the end of the year as a gift for the teacher to share with the next year's class to help acclimate them to Kindergarten.

NOTE This activity is similar to Activity 5-1, Order of Daily Events, in the *Teacher's Guide to Activities*.

Making Birthday Cards

Strand Number and Numeration

Focus Practice counting and number writing while creating birthday cards for classmates, friends, and family.

Children can design and make birthday cards for each other in the Art or Writing Center. Encourage children to decorate their cards with the appropriate number of candles on a cake or with other drawings that include the same number of objects as the child's age (such as five balloons). Have them label their pictures with the correct numeral. (Include models for writing each numeral in the Writing Center.) Keep the materials in the center to allow children to make birthday cards year-round. (In addition to crayons and paper, materials such as stickers, glue, construction paper, scissors, and ribbon will spark creativity.) Invite children to make birthday cards for family members or other friends, too.

Playing *I Have One, I Have Two*

> **Strand** Number and Numeration
>
> **Focus** Count and compare numbers related to children's bodies, clothing, or other personal information.

Children can take turns making number statements that describe something about themselves. For example, the first child might say, "I have 1 nose." The next child might say "I have 2 thumbs." The next child would say "I have 3 buttons on my shirt." Children can continue adding one at each turn until everyone has had a turn. Or children can continue to a predetermined number (such as 10), and then reverse the order or start again with 1 on the next child's turn.

Making "All About Me" Books and Timelines

> **Strands** Number and Numeration; Measurement and Reference Frames
>
> **Focus** Use numbers to describe information about themselves.

As children make "All About Me" books, they can incorporate mathematics by including information such as their address and phone number and personal measurements (height, weight, head circumference, arm span, and so on). They can also include other numeric data, such as the number of teeth they have (or have lost), number of people in their family, number of pets they have, and amount of money in their piggy banks. Children might also include a timeline of their lives to date using drawings or photographs to depict milestones and events from each year of their lives.

NOTE See the Teaching Options in Activity 5-1 (Order of Daily Events) of the *Teacher's Guide to Activities* for suggestions about making timelines of children's lives.

Animals All Around

Children see animals around them every day. These activities involve children in using mathematics as they investigate animals of all types. They might compare animal sizes and shapes, count and graph votes about favorite animals, sort animals by characteristics, and make animal pictures and sculptures out of 2- and 3-dimensional shapes.

Favorite Animal Graph

Strand Data and Chance

Focus Create and analyze a graph about favorite animals.

Have children draw a picture of their favorite animal on a stick-on note or small index card. Then create a bar graph with the animal drawings. Involve children by asking questions such as: *Which way should the bars go? How should we label the sides of the graph? What should the title be?* When the graph is complete, children should count and compare the number of each type of animal. Encourage children to draw conclusions from the graph. You might want to ask questions such as: *Which animal is the favorite? Which animal is the least favorite? How many more people chose zebras than lions?*

NOTE Children make bar graphs in Activities 1-8 (Birthday Graphs), 3-14 (Favorite Colors Graph), and 5-13 (Pet Bar Graph), among others. Refer to those activities for more specific suggestions about creating class bar graphs.

Sorting Animals

Strand Patterns, Functions, and Algebra

Focus Practice sorting animals in different ways.

Cut pictures of animals from magazines and glue them on large index cards or construction paper. (If you prefer, children can cut out or draw pictures of animals.) Encourage children to sort the animal pictures in different ways and explain how they sorted. (They might create labels for their sorting groups.) For example, children can sort by number of legs; spots, stripes, or other markings; size; places where the animals live (jungle, desert, farm); and so on. Thinking about the various ways that animals are alike and different helps children attend to details and, eventually, to better understand animal families and other categories of animals (mammals, reptiles, and so on). Invite children to share their sorting schemes with the group. (Children can also sort toy animals, if you have them in your classroom.)

Animal Patterns

Strand Patterns, Functions, and Algebra

Focus Identify and describe patterns in animal markings.

Using animal picture cards (see the previous activity, "Sorting Animals," for ideas about making these cards), have children describe patterns on the markings of some animals (for example, a black and white stripe pattern on a zebra). Discuss why certain animals might have certain markings (to blend in with their surroundings). You might want to provide paint and paper for children to replicate the animal patterns. Children can also search nature magazines and nonfiction books to find animals with patterned markings.

What Animal Am I?

Strand Patterns, Functions, and Algebra

Focus Identify animals using multiple attributes.

Have one child think of an animal and the other children in the group try to guess the animal by asking questions that can be answered with *yes* or *no*. For example, *Does the animal have spots? Does it have stripes? Does it live in a jungle?* Children take turns asking questions until they guess the correct animal. If children have difficulty thinking of questions, encourage them to think about specific features of the animals such as color, number of legs, or distinctive body parts like wings or a tail. This game provides an excellent opportunity for children to sharpen their problem-solving skills and to focus on attributes and sorting rules.

NOTE This game is similar to *Read My Mind,* which is played with attribute blocks in Activity 6-12.

Comparing Animal Sizes

Strand Measurement and Reference Frames

Focus Explore and compare the lengths of different animals.

Use reference books (or the Internet) to gather and record size information about various animals that interest children. For example, boa constrictor snakes can grow as long as 18 feet. The smallest hummingbird is only 3 inches long. Cut lengths of string that match the length of each animal (children can help with this task). Children can directly compare the lengths of the strings and describe their findings with measurement comparison words, such as *longer, shorter, longest,* and so on. They can also measure the strings with other measuring tools and units, such as blocks, connecting cubes, or children's feet and then compare these measurements. If real animals are available, help children make approximate measurements, being mindful to handle the animals gently.

Similarly, children can explore and compare the weights of different animals by loading items into a bag on a bathroom scale to match various animals' weights.

NOTE You might read *Actual Size* by Steve Jenkins (Houghton Mifflin, 2004), which uses pictures in interesting ways to compare the sizes of different animals.

Animal Shape Pictures

Strands Number and Numeration; Geometry

Focus Create animals using 2-dimensional shapes.

Children can use various geometric shapes to make animals. You might provide paper shapes (different colors and sizes) for gluing, or sponges cut into different shapes (and sizes) for dipping into paint and printing. Encourage children to name and count the different shapes they used to make the animals. *Color Zoo* by Lois Ehlert (HarperCollins, 1989) might spark children's ideas for this project. Children can also write or dictate a story about their animal when they are finished. The pictures and stories can be displayed together or compiled into a class book.

Crazy 3-D Creatures

Strands Number and Numeration; Geometry

Focus Construct animals using 3-dimensional shapes.

Collect boxes and various recyclable materials that children can use to make 3-dimensional animal sculptures. Invite children to use the materials to construct real or imaginary animals. Encourage children to count and describe the different parts they add to the animals (such as the number of teeth or legs) and to talk about the shapes of the body parts. Informally use the names of 2- and 3-dimensional shapes as you talk with children about their sculptures. For example, *Are you going to use this cube* (a box) *for the body? You used a cylinder* (a paper towel tube) *to make the neck. You drew triangles for eyes.* When the sculptures are completed, display them. Children can describe their creations and have a partner or the group try to guess which sculpture is being described. Children can also write or dictate stories to display with their animals.

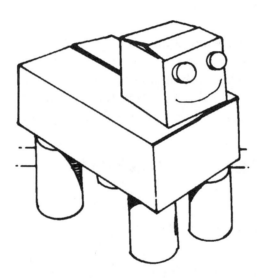

Dinosaurs

Children are curious about what types of dinosaurs existed, what they looked like, and how large they were. Exploring dinosaurs offers numerous opportunities to integrate math in interesting ways, such as looking at attributes, sorting into groups, and measuring and comparing sizes.

Sorting Dinosaurs

Strand Patterns, Functions, and Algebra

Focus Sort dinosaurs in various ways.

Provide an assortment of small dinosaur toys, models, or pictures of dinosaurs for children to sort. Encourage them to find different ways to sort and to explain (and perhaps label) their categories. You might keep a running list of all the ways children sort the dinosaurs (for example: by size, type of food they ate, if they walked on 2 or 4 legs, or other characteristics you have discussed).

Pattern-Block Dinosaurs

Strand Geometry

Focus Construct dinosaurs using pattern-block shapes.

Once children are familiar with pattern blocks, have them use the pattern blocks to construct small models of dinosaurs. Children may make realistic-looking dinosaurs or dinosaurs of their own creation. Encourage them to describe their dinosaurs and the different shapes or patterns they used in their constructions. Children could trace around the pattern blocks with a pencil or use the Pattern-Block Template to create a permanent record of their dinosaurs.

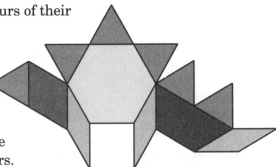

What Fits in a Dinosaur Footprint?

Strands Number and Numeration; Measurement and Reference Frames

Focus Explore measurement by figuring out how many things will fit in a dinosaur footprint.

Use masking tape to make the outline of a dinosaur footprint on the floor. The footprint should be about 2 feet long and $2\frac{1}{2}$ feet wide. This is the size of the footprint of one of the long-necked, long-tailed, plant-eating types of dinosaurs called *sauropods*. Ask children to figure out (by trial and error) how many children fit in the footprint and whether the number changes if the children are standing, sitting, or lying down. Help them record and compare their answers.

Ask children to think of other things that could fit into the footprint, then estimate and figure out how many would fit. Some possible objects are chairs, books, paper plates, or blocks. Children should compare and discuss their findings.

How Many Teeth Long?

Strand Measurement and Reference Frames

Focus Practice measuring with nonstandard units.

Make enough copies of Theme Master 1, page 80, so all children will have a tooth. Cut out the teeth and laminate them or mount them on heavy paper. Tell children that this is about the size of a tooth of an adult Tyrannosaurus Rex. Have children work together to measure something in the room by lining up their dinosaur teeth end to end, alternating and counting as they move forward. If measuring on their own, children should mark with a finger where the front of the tooth ends, then pick up and place the end of the tooth at that point. Be sure children keep track of how many times the tooth is put down. Children should record their results as: ___ dinosaur teeth.

NOTE Children measure with their own body parts in several activities in Section 5 of the *Teacher's Guide to Activities*.

Favorite Dinosaur Graph

Strands Number and Numeration; Data and Chance

Focus Create and analyze a graph about favorite dinosaurs.

With children, make a list of dinosaurs that the class has learned about. Have children draw (or in some way represent) a picture of their favorite dinosaur. Involve them in deciding how to use their drawings to create a bar graph of the class's favorite dinosaurs. (*Which way should the bars go? How should we label the sides of the graph? What should the title be?*) When the graph is complete, children should count how many children chose each dinosaur as their favorite and compare the numbers. Encourage children to answer (and ask) other questions about the graph, too. For example, ask: *How many more people chose Tyrannosaurus Rex than Pteranodon? How many chose plant eaters?*

Alternately, the class can use the list of dinosaurs as the basis for a tally chart of their favorite dinosaurs.

NOTE Children make bar graphs in Activities 1-8 (Birthday Graphs), 3-14 (Favorite Colors Graph), and 5-13 (Pet Bar Graph), as well as several others. You might want to refer to those activities for more specific suggestions about creating class bar graphs. Children make tally charts in Activity 5-9.

Dinosaur Puzzles

Strand Geometry

Focus Use spatial relations to assemble dinosaur puzzles.

Make 2 to 4 copies of the dinosaurs on Theme Master 2, page 81, and mount the copies on cardstock or heavy paper. Use different-colored paper as the background for each dinosaur, or have children color each dinosaur a different color. Laminate the dinosaurs, if possible. Cut each dinosaur into thirds, using the notch marks on the pictures as guides. Hide the pieces in the sandbox or sand table. Children can act as paleontologists to unearth the pieces and reassemble them using color and shape as a guide. (Explain to children that paleontologists actually dig to unearth dinosaur bones!) Watch whether children are comfortable flipping or rotating the pieces and placing them in the correct position relative to each other.

You can also use the dinosaur pictures on Theme Master 2 to make a memory game for partners or small groups. Copy the dinosaur pictures and cut each one in half (rather than thirds). Place the cards facedown. Players take turns turning over two cards. If they get two parts of the same dinosaur, they keep the cards. If not, they turn them back over. Play continues until all of the matches have been found.

Save the Plant Eater Game

Strand Number and Numeration

Focus Practice counting and using the + and − symbols in the context of a game.

In this board game, partners or small groups of children collaborate to save a plant-eating dinosaur from a Tyrannosaurus Rex. Prepare the game as follows:

▷ Have children decorate the gameboard on Theme Master 3, page 82, to look like an environment where dinosaurs might have lived. (Provide some picture books for children to use as models.) Have children draw a safe place for a dinosaur (such as a forest or cave) at the end point.

▷ Label 2 blank dice. Mark the first with +1, +2, +3, +4, −1, and −2. Color the second green on four sides and red on two sides.

▷ Gather 2 game markers (one for a Tyrannosaurus and one for a plant-eating dinosaur—Apatosaurus, for example). You might use small dinosaur figures as markers.

Materials
- ☐ decorated gameboard from Theme Master 3
- ☐ 1 color die; 1 number die (as described above)
- ☐ 2 markers (1 Tyrannosaurus, 1 plant-eating dinosaur)

Players 2 or more

Skill Adding and subtracting

Object of the Game Move the plant eater to the end of the board.

Directions

1. Players take turns rolling both dice. If the **color die** shows red, the Tyrannosaurus moves forward or backward the number of spaces shown on the **number die** (forward for +; backward for −). If the color die shows green, the plant-eating dinosaur moves forward or backward according to the number die.

2. The game ends when the Tyrannosaurus lands on the space with the plant eater OR when the plant eater gets to the end of the board (safely).

After children have played several times, they might want to try other combinations on the dice that give a better advantage to either the plant-eating dinosaur or to the Tyrannosaurus Rex. Children might also want to give the plant eater a head start.

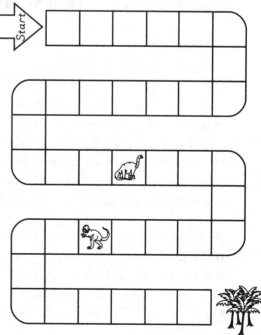

Forest

How Long Were the Dinosaurs?

Strand   Measurement and Reference Frames

Focus   Use nonstandard measurement units to explore and compare lengths of dinosaurs.

Have children choose several of their favorite dinosaurs and compare their lengths using children's outstretched arms as a unit of measure. You will need a long space, such as a hallway, if comparing very large dinosaurs. Look up the length of several dinosaurs, including some that are very large and some that are much smaller. For each dinosaur, show the distance from the head to the tail using a length of string or "head" and "tail" marks on the floor or wall. (You can use tape or lines to indicate the head and tail, or children can draw a head and tail.) One by one, call children to stand side by side with their arms outstretched and fingers touching until they match the length between the head and the tail. Record the measurement as: ___ children. If the number of children is not exact, decide as a group how to record the number. As you repeat for various dinosaurs, invite children to estimate about how many children will be in the line, using the results from previous dinosaurs as a reference. Compare the results after you've measured. Engage children in finding an interesting way to display the results, perhaps by making a concrete "graph" with lengths of string in the hallway or by creating a chart with a picture of each dinosaur and its size in "children units."

Dinosaur, Dinosaur, Where's Your Egg?

Strand   Patterns, Functions, and Algebra

Focus   Use attributes to identify the bone "thief."

This activity is a version of "Doggie, Doggie, Where's Your Bone?" Children sit in a circle. Choose one child to be the dinosaur who puts the egg in front of her and closes her eyes. Choose another child to take the egg and hide it behind his back. All children put their hands behind their backs and say: *Dinosaur, dinosaur, where's your egg?* The dinosaur opens her eyes and asks *yes* or *no* questions about the attributes of the person holding the egg, such as: *Is it a girl? Is it someone wearing blue?* The dinosaur uses the responses to guess who took the egg. (In some classrooms, the dinosaur is limited to 5 questions. If the dinosaur does not figure out who took the egg after 5 questions, he or she can call on children to give additional attribute-based hints.) To incorporate dinosaur facts that children have learned, allow the dinosaur to say what kind of dinosaur she is and share some information about that kind of dinosaur.

Fairy Tales

Literature based themes using fairy tales, nursery rhymes, and folk tales offer rich opportunities to connect language arts and mathematics. Activities that include mapping, graphing, sorting, patterning, and counting enrich the stories and help to develop mathematics skills at the same time.

Sorting Characters in Stories

Strands Data and Chance; Patterns, Functions, and Algebra

Focus Identify and sort characters using a variety of attributes.

Have children make a paper doll of a favorite character from a fairy tale or other story you have read. Provide cutout paper doll outlines from Theme Master 4, page 83, or let children make their own. Provide a variety of materials for children to use to dress their character such as fabric scraps, foil, paper, glue, markers, beads, buttons, and sequins. Talk with children about the types of characters that are in the fairy tales they have read. For example, they might mention kings and queens, princesses and princes, bad guys and good guys, girls and boys, characters who do magic, animals, or children. As a group, sort children's paper dolls in various ways according to the categories they suggest. You might make a picture graph that shows how many dolls you have in each category. Discuss whether characters can belong in more than one category. Gather children's ideas for showing this on the graph.

The class might be interested in setting up a tally chart to track the number of characters of each type for the fairy tales you have read. Each time you read a new fairy tale, the class can add tally marks to the chart.

Making Story Timelines

Strands Number and Numeration; Measurement and Reference Frames

Focus Sequence events in a story.

Children can make pictorial timelines that sequence the events of fairy tales or other stories. You might have children draw pictures of different events in the story, then help them order the pictures and label them with ordinal numbers (1st, 2nd, 3rd, and so on). Or, you might give them a paper divided into three parts and have them draw an event from the beginning of the story in the first section, an event from the middle of the story in the second section, and an event from the end of the story in the final section.

Making Maps from Stories

Strands Measurement and Reference Frames; Geometry

Focus Use measurement and spatial relations concepts to create maps from stories.

Children can create maps that give information about places or movement from place to place in stories. (These can be individual maps or small group or whole class collaborations.) To help children begin their maps, encourage them to think about the relationship of places and objects to each other in a story. For instance, ask questions such as:

▷ When Little Red Riding Hood travels to her Grandma's house, is it a short or a long distance? What does she walk through? What does she see as she travels?

▷ How does the Frog Prince go to visit the Princess? What rooms are in the castle? How many hops might be needed to get to the Princess's bedroom?

▷ How would you show the roads the Three Little Pigs take as they set out to see the world? Where are the pigs' houses?

If individual children make maps, have them share and explain the maps to each other. Create a map book to keep in class. *The Once Upon a Time Map Book,* by B.G. Hennessy (Candlewick, 2004) is an excellent complement to this activity (and to any aspect of a Fairy Tales Unit) and would inspire many mapping ideas.

NOTE In mapping, children use mathematical concepts, such as comparing distances and describing relative positions (right, left, next to, between, and so on). Several mapping activities are suggested in Project 6 in the *Teacher's Guide to Activities*.

Finding and Sorting Buried Treasure

Strands Number and Numeration; Patterns, Functions, and Algebra

Focus Practice counting and sorting buried treasure.

Many fairy tales involve hidden treasure! You might hide some "treasure" (such as fake jewels, gem-like rocks, coins, or buttons) in the sandbox or sand table. Let children dig and sift through the sand to find the treasure. Provide containers for children to count and record what they find. Encourage children to find different ways to sort the "treasure."

Making Props

Strands Measurement and Reference Frames; Geometry; Patterns, Functions, and Algebra

Focus Explore shapes, patterns, and measurement through an art project.

Children can create necklaces, bracelets, belts, crowns, or other props and treasures to use in acting out fairy tales. Provide yarn or pipe cleaners for stringing jewelry and long strips of paper for making belts and crowns. Also set out a variety of decorative materials, such as dyed macaroni, beads, buttons, and pieces of shiny paper or fabric. Encourage children to create and describe patterns as they make their props.

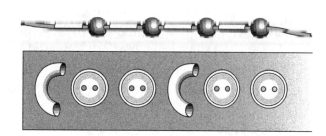

Discuss any shapes you notice in their projects. Children might measure to figure out what size belt, crown, or other prop they need.

Voting for Favorites

Strands Number and Numeration; Data and Chance

Focus Practice counting, tallying, and comparing numbers in the context of a class vote.

As a group, list characters in the stories that have been read. Ask children to vote for their favorite characters. Use tally marks to record the vote. Count the votes and compare the results. Children can also vote for favorite (and least favorite) stories; scariest stories; or other topics on which they might have different opinions.

Goldilocks	Papa Bear	Mama Bear	Baby Bear
//	///	////	⫰⫰ ///

Construct a Building

Strands Measurement and Reference Frames; Geometry

Focus Explore 3-dimensional shapes and measurement in the context of a building project.

Put out small boxes and tubes, lightweight cardboard, paper, glue, tape, markers, and scissors for children to use to construct structures from stories you have read (such as a castle or the houses in "The Three Little Pigs"). Encourage children to experiment with materials to create a sturdy base, to balance and support walls, and to determine correct sizes. Provide a ruler or measuring tape—as well as other available measuring tools—to encourage children to compare and measure lengths. Ask: *What shapes did you use to make the buildings? How tall is your structure? Do you want to weigh it?*

Treasure Squeeze

Strand Number and Numeration

Focus Use comparison clues to identify a number.

Set up a row of 10 upside-down paper or opaque plastic cups and label them in order from 1 through 10. Hide a treasure (such as a jewel) under one of the cups. Have children guess the cup the jewel is under. After a number is guessed, respond accordingly: *Your number is too large (or too small).* Continue until the correct number is guessed. Increase the number of cups when children are ready. Encourage them to play on their own with one child guessing and the other child hiding the treasure and responding with the clues *too large* or *too small.*

NOTE This game is an adaptation of *Monster Squeeze,* which is introduced in Activity 3-6 in the *Teacher's Guide to Activities.*

Make a Fairy Tale Game

Strand Number and Numeration

Focus Practice counting and numeral recognition in the context of a board game.

Provide copies of the blank gameboard (Theme Master 3, page 82) for creating games based on stories. For example, children might create games called *The Tortoise and the Hare Race, Help Cinderella,* or *Dragon and Knight Race to the Castle.* Encourage children to decorate the master based on the themes of their games. Provide dice and markers. Children can use basic rules, such as taking turns rolling the die and moving forward that number of spaces. Or let children make up other rules, such as skipping a turn when a 6 is rolled, or moving backward 1 space when a 1 is rolled. They can also make their own dice. Provide time for children to play their games with each other.

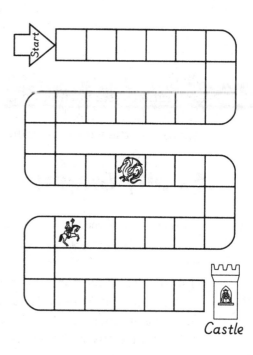

Castle

Is It Possible?

Strand Data and Chance

Focus Discuss probability of events in fairy tales.

Using a familiar fairy tale or nursery rhyme, list some of the things that happen in the story. Discuss these events and sort them into lists of "likely," "not likely," and "impossible." For example: *Is it possible to trade something to get something else? Is it possible to climb to the sky on a beanstalk?*

Families at Home

Children can do many mathematics-linked activities to further their exploration of similarities and differences among family sizes, family routines, and the places where families live. As you do these activities, remember to be sensitive to the variety of children's family and home situations.

Family Graphs

Strands Number and Numeration; Data and Chance

Focus Create and analyze a graph about family size.

Children can make pictures or write numerals to show the number of people in their families. Explain that families are groups of people who love and take care of each other, and that families are not all the same. (Children can define their families any way they like. Their definitions can include: all of the people who live with them, their parent(s) and siblings, or parents, siblings, aunts, uncles, cousins, and/or grandparents. Be prepared to address any questions or comments that may arise.) As a class, create a bar or picture graph representing the total number of people in each child's family. Compare the number of people in different families. Ask questions such as: *How many people in our class have four members in their family? How many people are in the biggest family on our graph?*

NOTE Children make bar graphs in Activities 1-8 (Birthday Graphs), 3-14 (Favorite Colors Graph), and 5-13 (Pet Bar Graph), as well as several others. You might want to refer to those activities for more specific suggestions about creating class bar graphs.

Brother and Sister Chart

> **Strands** Number and Numeration; Data and Chance
>
> **Focus** Create and analyze a graph to show numbers of sisters and brothers.

Make a chart to show how many brothers and sisters each child in the class has. Make a list of children's names in a single column on the left edge of a large piece of paper. Leave about two inches between each name so children can place index cards or stick-on notes in a row next to their names. Collect two colors of index cards or stick-on notes (one color to represent brothers; the other to represent sisters). Ask each child: *How many brothers do you have?* Have the child take a corresponding number of one of the colors of cards or stick-on notes. Then ask: *How many sisters do you have?* Have the child take the correct number of the other color cards or notes. Each child should place his or her cards or notes side by side in a row next to his or her name, starting with all of the "brother" cards on the left, then adding the "sister" cards. For children who do not have brothers or sisters, reinforce the concept of zero. When the chart is complete, begin by asking how many brothers and sisters each child has. Continue with other questions, including those that require children to consider brothers and sisters separately. This may be difficult for many! For example: *Who has the most brothers? Who has the most sisters? Which children have more sisters than brothers?*

NOTE This activity produces a different type of data display than the Family Bar Graph in the previous activity. It is good for children to work with different types of data displays.

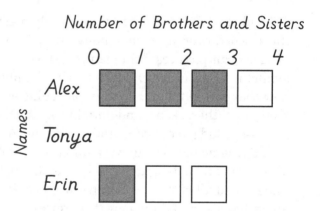

Skip Counting Family Features

> **Strand** Number and Numeration
>
> **Focus** Use body parts to practice skip counting.

Have children draw pictures of everyone in their families or create individual paper dolls for their family members. (Use the paper doll outlines on Theme Master 4, page 83, if desired.) Encourage children to add details such as eyes, toes, fingers, and so on. Have children use the pictures or dolls to practice skip counting. For example, you might prompt them to count the number of eyes in their families by 2s. Or you might ask: *What can we count by 5s or 10s? How many fingers (or toes, or fingers and toes) in your family?* Encourage children to think of other body features to count and ways to count them. Two or more children can join families to skip count to higher numbers.

"A Day at Home" Books or Timelines

Strand   Measurement and Reference Frames

Focus   Sequence daily events from children's lives.

Invite children to use sequencing skills to make a book or timeline about a day at home with their families. Encourage them to think of a recent day and draw pictures of at least three things they did during that day (on separate sheets of paper). Help them write captions for their pictures. Ask questions to help children sequence the events in order. For example, *What did you do first? What happened next? What was the last thing you did?* Collate each child's pictures into his or her own "A Day at Home" book, or display them as timelines.

NOTE This activity is similar to Activity 5-1, Order of Daily Events in the *Teacher's Guide to Activities.*

Family Sorting Cards

Strand   Patterns, Functions, and Algebra

Focus   Sort photos of family members in different ways.

Take photos of children and their family members at the beginning or end of the school day or at an open house, or ask children to bring photos from home. (You can cut pictures from magazines if photos from class families are not available.) Encourage children to sort the photos in different ways and explain how they sorted them. For example, children might sort pictures into groups of mothers, fathers, sons, and daughters. Or they might make groups of babies, school-age kids, teenagers, and grownups. As children sort, they may realize that the same person can have different family roles (for example, a brother is also a son, and a daughter can be a mother). Children do not need to understand all of these relationships, but they may be interested in talking about them and making new discoveries.

Building Homes

Strands Number and Numeration; Measurement and Reference Frames; Geometry

Focus To explore shape and measurement by building 2- or 3-dimensional models of homes.

Provide materials for children to represent the places where they live. For example, you might supply pre-cut paper shapes (or scissors to cut shapes) for making 2-dimensional collage representations. Or you might supply boxes in various shapes and sizes, cardboard tubes, recyclables, and other materials for children to make 3-dimensional structures. Encourage children to think about details such as the shape of the roof and the shapes and number of windows and doors. Help them write their addresses on their pictures or structures. (You can have them add phone numbers, too, if children are learning those.)

If you have a large shipping box (from kitchen appliances or furniture), children can use it to make a pretend home in the Dramatic Play Center. In addition to thinking about shapes and numbers of features, they might want to measure their model home or its components.

Where Do I Live?

Strands Number and Numeration; Data and Chance

Focus Sort and graph different types of homes.

Read *Houses and Homes* by Ann Morris (HarperTrophy, 1995), and discuss the idea that people live in different types of homes. Have children draw pictures of where they live. Decide with them how to sort and label the categories of homes so that all the types of homes in the class are included (house, apartment, and trailer, for example). Use the pictures to create a graph that shows how many children live in each type of home. Count, record, and compare the number of pictures in each category.

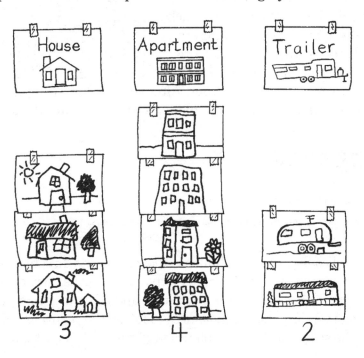

Theme 6: Families at Home **33**

Growing Things

As children observe and care for plants outdoors, in the classroom, or at home, they have opportunities to count, sort, measure, and make comparisons.

Measuring Plants

> **Strand** Measurement and Reference Frames
>
> **Focus** Compare and track heights of plants.

Each child can plant a daffodil, paper white narcissus, or onion bulb in a cup or flowerpot filled with soil. Label the container with the child's name. Tape a ruler (or a laminated copy of a ruler) to a craft stick so the craft stick extends below the zero end of the ruler. Insert the craft stick in the cup so that the zero on the ruler is at the surface of the soil. Add water according to the directions on the package of bulbs. As the plants grow, children measure the plants' heights by comparing the top of the plant to the inch marks on the ruler. You might want to have children record the measurements of their plants and the dates for each measurement. Encourage them to figure out how much their plant has grown between measurements.

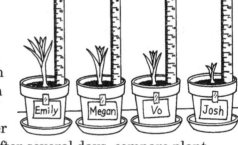

As the plants grow, periodically discuss and compare their heights. Have children figure out which plant is the tallest, which is the next tallest, and so on. Children can line up the plants in order from tallest to shortest. It may be helpful to use a smaller number of plants to compare and order. After several days, compare plant heights again. Check the order of the plants to see if they are the same.

Planting Seeds

> **Strands** Number and Numeration; Measurement and Reference Frames
>
> **Focus** Apply counting and measurement skills when planting seeds.

If you plant seeds outdoors with children, show them the directions on the seed packets that indicate how far apart to plant the seeds, how deep to dig the holes, how many seeds to put in each hole, and other planting guidelines. Use measuring tools or personal measures to follow the planting guidelines. Discuss why the seeds for different plants might have different guidelines.

Sorting Flowers and Seeds

Strand Patterns, Functions, and Algebra

Focus Sort flowers and seeds according to various attributes.

Cut photographs or pictures of flowers from magazines or seed catalogs. Invite children to sort the flowers in different ways and to explain their sorting rules. For example, children might sort by colors, by flower shape, or by number of petals. Children's sorting categories may provide natural opportunities to talk with them about why flowers have certain colors, shapes, or structures (to attract birds or insects for pollination, for example). You can also provide a variety of seeds for sorting.

NOTE Check whether any children have seed or nut allergies before working with seeds.

Estimating Seeds

Strands Number and Numeration; Measurement and Reference Frames

Focus Estimate, count, and compare seeds from different types of fruits.

Explain that fruits contain the seed(s) of a plant and that new plants grow from the seeds. Choose different types of fruits, such as apples, peaches, and pumpkins, and have children estimate how many seeds each one has. Have children count and then compare the numbers of seeds with their predictions. Record the number of seeds that are found for each fruit and compare and discuss these numbers, too. *(Why do pumpkins have so many more seeds than peaches? Have you ever seen a plum with two seeds? Could you count all of the seeds on the outside of the strawberry?)* Children can also compare the sizes and shapes of the seeds from the various types of fruits. Provide measuring tools, such as rulers, pan balances, and scales for measuring.

Seed Collages

> **Strands** Number and Numeration; Patterns, Functions, and Algebra
>
> **Focus** Use seeds for patterning, counting, and estimating.

Collect a variety of seeds for children to glue onto paper or cardboard to make seed collages. Encourage children to arrange the seeds in patterns on their collages. As they work, talk with them about the shapes and relative sizes of the seeds they are using. When they finish, you might also ask them to count how many of each type of seed they used (if the number isn't huge).

You might also provide flower outlines from Theme Master 5, page 84, for children to fill with seeds. Have them estimate how many of a particular kind of seed will fit, then test their estimates by gluing seeds onto the paper flowers.

How Many Seeds?

> **Strands** Number and Numeration; Operations and Computation
>
> **Focus** Practice problem-solving using seeds as counters.

Children can play a problem-solving game using seeds as counters. Children face each other with a row of 5 large seeds between them. Players count the seeds together to begin the game. Then, one player closes his or her eyes while the other takes between 1 and 4 seeds and hides them behind his or her back. The first player opens his or her eyes and tries to figure out how many seeds were taken. After the first player guesses, the second player lays the missing seeds down next to the remaining seeds in the middle, counting aloud as each seed is added. Adjust the number of seeds in the middle depending on the children's comfort level. Some children may be able to play with 10 or more seeds in the middle.

Busy Bee Game

Strand   Number and Numeration

Focus   Practice counting using a board game.

In this board game, partners or small groups of children move their bees to the matching flower at the end of the board. Prepare the game as follows:

▷ Use the blank gameboard (Theme Master 3, page 82), or draw a wavy path of 24 squares on posterboard, stopping about four inches from the end of the board. In this space, glue four small flowers from Theme Master 5 (page 84), and color the flowers blue, red, green, and yellow. Draw arrows that connect the last square on the gameboard to each flower. (Children can help prepare the gameboard.)

▷ Cut the bees from Theme Master 5 and color them to match the flowers on the board. Glue the bees to counters or bottle caps, if possible.

Materials   ☐ gameboard (as described above)
                   ☐ bees and flowers (as described above)
                   ☐ a die

Players   2 to 4

Skill   Counting

Object of the Game   Move the bee to the flower at the end of the path

Directions

1. To play, each child rolls the die and moves a bee marker that number of spaces along the path.

2. To land on the flower at the end, the player must roll the exact number it takes to reach the flower. Continue playing until all the bees reach their matching colored flowers.

You can make the game more challenging by customizing the die with addition and subtraction symbols (+1, +2, +3, +4, −1, −2, for example) or other variations.

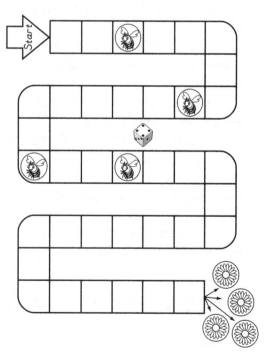

Seasons

Throughout the year, children observe changes in the weather and the seasons. These mathematics activities can be spaced to correspond with seasonal changes in your area.

Seasons Calendar

> **Strands** Measurement and Reference Frames; Patterns, Functions, and Algebra
>
> **Focus** Become familiar with seasonal events and the progression of seasons.

Label four pieces of posterboard "Fall," "Winter," "Spring," and "Summer." Ask children what things happen during each season (holidays, types of weather, activities), and have children draw pictures of things they associate with the seasons. Children can post their pictures on the appropriate posterboard. (Children can also use the pictures on Theme Master 6, page 85.) You might arrange the sheets of posterboard in order according to the school year and move an arrow to the correct posterboard at the start of each season. You might also add the names of the months for each season at the top of each posterboard. (Transition months can be listed for two seasons, or you might write "early June" and "late June," for example.) Invite children to add pictures to the posterboards at any time.

Temperature and the Seasons

> **Strand** Measurement and Reference Frames
>
> **Focus** Explore and compare seasonal temperature trends.

At the end of each season, look at your temperature display from Routine 7 (Recording Daily Temperature, page 28, of the *Teacher's Guide to Activities*.) Discuss temperature trends that you notice within the season and from the previous season. Children can make predictions about the temperature in the season ahead.

Seasonal Collections

Strands Number and Numeration; Patterns, Functions, and Algebra

Focus Sort seasonal objects in various ways.

Invite children to make seasonal nature collections (leaves in the fall, flowers in the spring, for example). Have children sort the items in the collection in different ways and explain how they sorted. They can also count the number of items in each group and the total number of items in the collection.

Seasonal Graphs

Strand Data and Chance

Focus Graph favorite seasons and seasonal activities.

At the beginning of each season, generate a list of things that children might do outdoors during that season. Then have them vote for their favorite activities on the list. You might record the votes with tally marks or display the results on a bar graph. You can also make a graph or tally chart that shows how many children choose each season as their favorite.

Puddle Prints

Strands Number and Numeration; Measurement and Reference Frames; Patterns, Functions, and Algebra

Focus Use wet footprints for patterning, measuring, and counting.

When outdoors in rain or snow, children can make wet footprints for measuring, counting, and creating patterns. (You can also use several pans with small amounts of water. Have children step into the pans, and then use their wet shoes to make prints on the sidewalk.) Children can measure the lengths of the wet footprints. You might suggest that they make lines of footprints and count how many footprints they can make before running out of water. They can work with a partner to make patterns, such as large footprint, small footprint, large footprint, small footprint, and so on.

NOTE Send a note home explaining the activity and requesting that children wear shoes with thick rubber soles or other shoes that will not be damaged by water. Children can also do this activity with bare feet in warm weather.

Measuring Rain and Snow

Strand Measurement and Reference Frames

Focus Figure out ways to measure the amount of rain or snowfall.

On a rainy or snowy day, help children put a bowl outside to collect rain or snow. After the rain or snow stops (or near the end of the day), bring the container inside and engage children in discussing ways to measure how much water was collected. Children might suggest pouring the water into a measuring cup or using a ruler to measure the height of the water level in the bowl. Make a record that describes your chosen measurement technique(s) and your results. Repeat during other rainstorms or snowfalls and compare the results. Each time, have children predict whether there will be more or less water than the last time you measured.

Exploring Raindrops

Strands Number and Numeration; Measurement and Reference Frames

Focus Explore volume and capacity using water and eyedroppers.

Give children eyedroppers and milk caps (or other very small containers) to use at the water table. (If you don't have a water table, you can use a bowl of water in a baking tray or dish tub to contain spills.) Encourage children to predict and then test how many drops of water will fit in the milk caps. Have them record their results. Ask them if they think the drops from the eyedropper are the same size as raindrops. How many raindrops do they think it takes to make a small puddle?

Exploring Snow

Strand Measurement and Reference Frames

Focus Compare the volume of frozen and melted snow.

Fill the water table (or dish tubs or baking pans) with clean snow from outdoors. Provide containers of different sizes and shapes and invite children to explore the snow. As children scoop and pack and pour the snow, discuss the sizes of the containers they are filling and compare the amount of snow that fits in the different containers and scoops. As the snow melts, encourage discussion about whether melted snow or frozen snow fills more space in the containers. Encourage children to test their theories by marking the level in a bowl of snow, then setting it out to melt and comparing the level of water in the container afterwards.

If snow is not available, make "pretend snow" by spraying several cans of shaving cream into the water table or bowls.

Paper Snowflakes

Strands Geometry; Patterns, Functions, and Algebra

Focus Investigate symmetry, shapes, and patterns through an art activity.

Demonstrate how to make paper snowflakes by folding a sheet of paper into quarters and cutting small shapes out of each side. (Children may need reminders not to cut the connecting folds all the way through.) When unfolded, the paper will have a cut design that resembles a snowflake. Encourage children to identify the shapes and patterns they have created and to consider whether their snowflakes are symmetrical. *The Snowy Day* by Ezra Jack Keats (Viking Juvenile, 1962) is a wonderful introduction to this activity.

NOTE Children make paper snowflakes to revisit the concept of symmetry in Activity 4-9 in the *Teacher's Guide to Activities*.

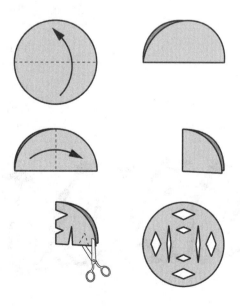

Songs and Poems

Counting

Ten Little Penguins

(Suggested in Activity 2-6)

Ten little penguins went waddling into town.
Along came the wind, whooooo *(blow)*
And knocked one down.
"Oh my," said the rest,
"We must do our best to keep from falling down."

(Repeat with nine little penguins, eight little penguins, and so on until two are left:)

Two little penguins went waddling into town.
Along came the wind, whooooo
And knocked one down.
"Oh my," said the last.
"I must do my best to keep from falling down."

One little penguin went waddling into town.
Along came the wind, whooooo
And knocked it down.

This Old Man

(Suggested in Activity 1-3)

This old man, he played one,
He played knick-knack on my thumb,

REFRAIN:
With a knick-knack paddywhack,
Give a dog a bone,
This old man came rolling home.

This old man, he played two,
He played knick-knack on my shoe.

(Add refrain after each verse.)

This old man, he played three,
   ... on my knee.
This old man, he played four,
   ... at my door.
This old man, he played five,
   ... on his tie.
This old man, he played six,
   ... on some sticks.
This old man, he played seven,
   ... up to heaven.
This old man, he played eight,
   ... at my gate.
This old man, he played nine,
   ... on a dime.
This old man, he played ten,
   ... once again.

With a knick-knack paddywhack,
Give a dog a bone,
This old man came rolling home.

Over in the Meadow

Over in the meadow,
In the sand in the sun
Lived an old mother toadie
And her little toadie one.
"Wink!" said the mother;
"I wink!" said the one.
So they winked and they blinked
In the sand in the sun

Over in the meadow,
Where the stream runs blue
Lived an old mother fish
And her little fishes two.
"Swim!" said the mother;
"We swim!" said the two.
So they swam and they leaped
Where the stream runs blue.

Over in the meadow,
In a hole in a tree
Lived an old mother bluebird
And her little birdies three.
"Sing!" said the mother;
"We sing!" said the three.
So they sang and were glad
In a hole in the tree.

Over in the meadow,
In the reeds on the shore
Lived an old mother muskrat
And her little ratties four.
"Dive!" said the mother;
"We dive!" said the four.
So they dived and they burrowed
In the reeds on the shore.

Over in the meadow,
In a snug beehive
Lived a mother honeybee
And her little bees five.
"Buzz!" said the mother;

"We buzz!" said the five.
So they buzzed and they hummed
In the snug beehive.

Over in the meadow,
In a nest built of sticks
Lived a black mother crow
And her little crows six.
"Caw!" said the mother;
"We caw!" said the six.
So they cawed and they called
In their nest built of sticks.

Over in the meadow,
Where the grass is so even
Lived a gay mother cricket
And her little crickets seven.
"Chirp!" said the mother;
"We chirp!" said the seven.
So they chirped cheery notes
In the grass soft and even.

Over in the meadow,
By the old mossy gate
Lived a brown mother lizard
And her little lizards eight.
"Bask!" said the mother;
"We bask!" said the eight.
So they basked in the sun
On the old mossy gate.

Over in the meadow,
Where the quiet pools shine
Lived a green mother frog
And her little froggies nine.
"Croak!" said the mother;
"We croak!" said the nine.
So they croaked and they splashed
Where the quiet pools shine.

Over in the meadow,
In a sly little den
Lived a gray mother spider
And her little spiders ten.
"Spin!" said the mother;
"We spin!" said the ten.
So they spun lacy webs
In their sly little den.

Five Little Ducks

Five little ducks went out to play,
Over the hill and far away.
When mother duck went *quack, quack, quack,*
Four little ducks came waddling back.

(Repeat with four little ducks, three little ducks,
two little ducks, and then continue:)

One little duck went out to play,
Over the hill and far away.
When mother duck went *quack, quack, quack,*
No little ducks came waddling back.
Sad mother duck went out one day,
Over the hill and far away.
Then sad mother duck went QUACK!
    QUACK! QUACK!
And five little ducks came waddling back.

Five Speckled Frogs

Five little speckled frogs, sat on a speckled log,
Eating the most delicious bugs, yum, yum yum!
One frog jumped in the pool,
Where it was nice and cool,
Now there are four little speckled frogs.
Glub, glub, glub!

(Repeat, counting down until there are none.)

Five Little Monkeys

(Suggested in Activity 1-4)
Five little monkeys jumping on the bed,
One fell off and bumped his head.
Mama called the doctor and the doctor said,
"No more monkeys jumping on the bed."

(Repeat, counting down until there are none.)

Countdown

Here are the astronauts in their
    shiny suits.
They climb in their capsule.
Now it's time to move.
10-9-8-7-6-5-4-3-2-1 BLAST OFF!

Calendar

Months of the Year

In January falls the snow.
In February cold winds blow.
March brings out the early flowers.
April brings the sunny showers.
In May the roses bloom so gay.
In June the farmer cuts his hay.
In July brightly shines the sun.
In August harvest has begun.
September turns the green leaves brown.
October winds then blow them down.
November days are dark and drear.
December comes and ends the year.

Macarena Months

January, February,
(Hands out with palms facing down.)
March, April,
(Turn hands over with palms facing up.)
May, June,
(Cross hands as they touch each shoulder.)
July, August,
(Uncross hands and move to back of head.)
September, October,
(Cross hands and place on thighs.)
November, December,
(Uncross hands and place on waist.)
Then you turn around.
(Repeat three more times and end with:)
Then you sit right down!

Days of the Week

(Tune: The Bear Went Over the Mountain)
There are seven days in the week,
Seven days in the week, seven days
  in the week,
And I can say them all!
Sunday, Monday and Tuesday,
Wednesday, Thursday and Friday
Saturday is the last day
And I can say them all!

Today Is Monday

Today is Monday, today is Monday.
Monday, pizza,
All you hungry children
Come and eat it up.

Today is Tuesday, today is Tuesday.
Tuesday, spaghetti
Monday, pizza,
All you hungry children
Come and eat it up.

Today is Wednesday, today is Wednesday.
Wednesday, SOOOOOOP
Tuesday, spaghetti
Monday, pizza,
All you hungry children
Come and eat it up.

*(Continue adding days and foods
until you reach Sunday. Children
can choose the foods to match
what they might eat that day.)*

Spatial Relationships

Going on a Bear Hunt

(Suggested in Activity 2-3)
(Children repeat each line after you.)

CHORUS:
*(Slap thighs during chorus
to set a walking beat.)*
Going on a bear hunt
I'm not afraid
Got a real good friend *(children hug)*
By my side.
Oh, Oh,
What do I see?

Oh look! It's some tall grass!
Can't go over it
Can't go under it
Can't go around it
Got to go through it.
(Pretend to clear a way through grass.)

CHORUS

Oh look! It's a tall tree.
Can't go over it
Can't go under it
Can't go through it
Got to climb up it.
(Pretend to climb up the tree.)

CHORUS

Oh look! It's a wide river.
Can't go over it
Can't go under it
Can't go through it
Got to swim across it.
(Pretend to swim.)

CHORUS

Oh look! A deep, dark cave
Can't go over it
Can't go under it
Can't go through it
Got to go in it.
(Close eyes and pretend to enter the cave.)

Oh, oh! It's dark in here.
I feel something
It has lots of hair!
It has sharp teeth!
It's a bear!!
*(At this point, chant faster and
reverse the order of events in the song.)*

Go back out of the cave,
Swim back across the river,
Go back up and down the tree,
Go back through the tall grass
Until you get safely home and lock the door.

I'm not afraid!

Go In and Out the Window

Go in and out the window.
Go in and out the window.
Go in and out the window, as we have done
    before.
*(Variations: Change the verse to add
other directions such as go round and
round, stand in front and back,
and so on.)*

Hokey Pokey

You put your left foot in,
You put your left foot out.
You put your left foot in,
And you shake it all about.
You do the Hokey Pokey
And you turn yourself about.
That's what it's all about.
*(Repeat with different parts of the
body, until "your whole self" is added.)*

Patterns

Did You Ever See a Lassie?

(Suggested in Activity 7-15)
*Use "lassie" or "laddie," depending on
whether a girl or a boy is acting out a
motion pattern "this way and that way."*

Did you ever see a lassie, a lassie, a lassie,
Did you ever see a lassie,
Go this way and that?
Go this way and that way,
And that way and this way,
Did you ever see a lassie
Go this way and that?

B-I-N-G-O

There was a farmer who had a dog,
And Bingo was his name-o.
B-I-N-G-O,
B-I-N-G-O,
B-I-N-G-O
And Bingo was his name-o.
*(Repeat, replacing a
successive letter of B-I-N-G-O
with a clap at each verse.)*

Money

Dollar Holler Rap

CHORUS:
Pennies, nickels, dimes, and quarters
Pennies, nickels, dimes, and quarters

My sweetie knows a song.
It's really kind of funny.
It's all about coins—
And learning about money.

CHORUS

Now a penny means 1   *(Point 1.)*
And a nickel means 5.   *(Flash 5.)*
Dimes are worth 10.   *(Show 10.)*
And quarters 25. *(Flash hands to show 25.)*

CHORUS

(Optional verse for coin exchanges)

Five pennies in a nickel,
Two nickels in a dime,
Five nickels in a quarter,
You'll know it every time.

CHORUS

Penny, Penny

Penny, penny, easily spent,
Copper brown and worth one cent.

Nickel, nickel, thick and fat,
It's worth five cents, I know that.

Dime, dime, little and thin,
I remember, you're worth ten.

Quarter, quarter, big and bold,
It's worth twenty-five, I am told!

Coin Rhyme

(Tune: Miss Lucy)
Five pennies make a nickel.
Two nickels make a dime.
Two dimes and a nickel make
   a quarter everytime.
Four quarters make a dollar
And that is quite a lot.
A dollar is exactly how much money
I have got!

Weather

Weather Song

(Tune: Row, Row, Row Your Boat)
Sun, wind, rain, or snow,
What will come our way?
Let's ask the meteorologist
About the weather today!

Weather Response Song

(Tune: BINGO)
There was a day and it was <u>hot</u> *(warm, cold)*
And <u>sunny</u> *(cloudy, rainy, snowy, foggy)* was the
   weather.
S-U-N-N-Y,
S-U-N-N-Y,
S-U-N-N-Y
And sunny was the weather.
*(Replace "hot" and "sunny" with the
weather for the day.)*

Number Writing

Air Writing Song

(Tune: If You're Happy & You Know It)
Put your finger in the air, in the air.
Put your finger in the air, in the air.
Put your finger in the air, and write
   the number ____.
Put your finger in the air, in the air.

The Numeral Song

(Tune: Skip to My Lou)
Come right down and that is all.
Come right down and that is all.
Come right down and that is all.
To make the numeral 1.
(Repeat, changing directions for each numeral.)
Curve around and slide to the right ... 2.
Curve in and around again ... 3.
Down, slide, cut it in two ... 4.
Across and down, then around ... 5.
Come right down and give it a curl ... 6.
Slide to the right and slant right down ... 7.
Make an "S" and close the gate ... 8.
Circle around and come right down ... 9.
Come right down then make a 0 (zero) ... 10.
We can sing the numeral song,
We can sing the numeral song,
We can sing the numeral song,
And make numerals all day long!!!

Operations

Addition and Subtraction Chants

Five little cookies in the bakery shop,
Shining bright with sugar on top.
Along comes (child's name)
With cookies today
And puts ____ more cookies on the tray.
How many cookies now?

Five pumpkin pies in the bakery shop,
Round and orange
With whipped cream on top.
Along comes (child's name)
With money to pay
And takes ____ pumpkin pies away!
How many pies now?

Literature Lists

Number and Numeration

Counting

Author	Title	Publisher	Program Link
Anno, Mitsumasa	*Anno's Counting Book*	HarperCollins, 1977	
Appelt, Kathi	*Bat Jamboree*	HarperCollins, 1998	Activity 2-9
Bajaj, Varsha	*How Many Kisses Do You Want Tonight?*	Little Brown Co., 2004	
Bang, Molly	*Ten, Nine, Eight*	William Morrow & Co., 1983	
Berkes, Marianne	*Over in the Ocean: In a Coral Reef*	Dawn Publications, 2004	
Christelow, Eileen	*Five Little Monkeys Jumping on the Bed*	Houghton Mifflin Harcourt, 1998	Activity 1-4
Dahl, Michael	*Counting by Fives*	Picture Window Books, 2004	Activity 5-8
Crews, Donald	*Ten Black Dots*	William Morrow & Co., 1995	
Dunbar, Joyce	*Ten Little Mice*	Gulliver Books, 1990	
Edwards, Pamela Duncan	*Warthogs in the Kitchen*	Hyperion, 1998	
Ehlert, Lois	*Fish Eyes: A Book You Can Count On*	Harcourt Brace, 1990	
Evans, Lezlie	*Can You Count Ten Toes? Count to 10 in 10 Different Languages*	Houghton Mifflin, 1999	
Feelings, Muriel	*Moja Means One: A Swahili Counting Book*	Dial Press, 1996	
Fleming, Denise	*Count!*	Henry Holt and Co., 1992	
Friedman, Aileen	*The King's Commissioners*	Scholastic 1995	Activity 6-14
Grossman, Bill	*My Little Sister Ate One Hare*	Crown Books, 1996	
Hamm, Diane Johnston	*How Many Feet in the Bed?*	Aladdin, 1994	Activity 6-10
Hoban, Tana	*Let's Count*	Greenwillow, 1999	
Hong, Lily Toy	*Two of Everything: A Chinese Folktale*	Albert Whitman & Co., 1993	
Hutchins, Pat	*Ten Red Apples*	Greenwillow, 2000	

Counting

Author	Title	Publisher	Program Link
Krebs, Laurie	We All Went on a Safari: A Counting Journey through Tanzania	Barefoot Books, 2003	
MacDonald, Suse	Look Whooo's Counting	Scholastic, 2000	
Martin, Bill	Chicka, Chicka 1, 2, 3	Simon & Schuster, 2004	Activity 5-4
Miranda, Anne	Monster Math	Harcourt, 1999	
Root, Phyllis	One Duck Stuck	Candlewick, 1998	
Ryan, Pam Muñoz	One Hundred Is a Family	Hyperion, 1996	Activity 3-15
Sarfatti, Esther	Counting by: Fives	Rourke, 2007	Activity 5-8
Sloat, Teri	From One to One Hundred	Puffin Books, 1995	Project 5
Stickland, Paul	Ten Terrible Dinosaurs	Dutton Children's Books, 1997	Dinosaurs Theme
Walsh, Ellen Stoll	Mouse Count	Harcourt, 1995	Activity 2-14
Wise, William	Ten Sly Piranhas	Puffin, 1993	

Big Numbers

Author	Title	Publisher	Program Link
Anno, Mitsumasa	Anno's Mysterious Multiplying Jar	Putnam, 1983	
Gag, Wanda	Millions of Cats	Putnam, 2004	
McKissack, Patricia C.	A Million Fish. . .More or Less	Random House, 1996	
Schwartz, David M.	How Much Is a Million?	HarperTrophy, 1993	

100th Day

Author	Title	Publisher	Program Link
Cuyler, Margery	100th Day Worries	Simon & Schuster, 1999	Project 5
Franco, Betsy	Counting Our Way to the 100th Day	Margaret K. McElderry, 2004	Project 5
Harris, Trudy	100 Days of School	Millbrook, 1999	Project 5
Haugen, Brenda	The 100th Day of School	Picture Window Books, 2003	Project 5
McCourt, Lisa	100th Day of Bug School	HarperFestival, 2004	Project 5
McMullan, Kate	Fluffy's 100th Day of School	Cartwheel, 2000	Project 5
Medearis, Angela Shelf	The 100th Day of School	Scholastic, 2005	Project 5
Schiller, Melissa	100th Day of School	Children's Press, 2003	Project 5
Slate, Joseph	Miss Bindergarten Celebrates the 100th Day of Kindergarten	Penguin Group, 2002	Project 5
Wells, Rosemary	Emily's First 100 Days of School	Hyperion, 2005	Activity 1-5 Project 5

Other Numeration Books

Author	Title	Publisher	Program Link
Alda, Arlene	*Arlene Alda's 1 2 3*	Ten Speed Press, 2004	Project 1
Burns, Marilyn	*Spaghetti and Meatballs for All: A Mathematical Story*	Scholastic Press, 2008	Project 4
Dee, Ruby	*Two Ways to Count to Ten: A Liberian Folktale*	Henry Holt & Co., 1990	Activity 6-10
Giganti, Paul, Jr.	*Each Orange Had 8 Slices*	Greenwillow Books, 1992	
Hutchins, Pat	*The Doorbell Rang*	HarperTrophy, 1999	Activity 6-11
Mahy, Margaret	*17 Kings and 42 Elephants*	Dial, 1987	
Murphy, Stuart	*Give Me Half!*	Steck-Vaughn, 1999	Activity 6-16
Oughton, Jerrie	*How the Stars Fell into the Sky: A Navajo Legend*	Houghton Mifflin Harcourt, 1996	Activity 5-15
Pinczes, Elinor J.	*One Hundred Hungry Ants*	Houghton Mifflin Harcourt, 1999	Project 5
Ross, Tony	*Centipede's One Hundred Shoes*	Henry Holt and Co., 2003	Project 5
Sayre, April Pulley and Jeff	*One Is a Snail, Ten Is a Crab*	Candlewick, 2006	Activity 8-9
Schlein, Miriam	*More Than One*	Greenwillow, 1996	
Schuette, Sarah L.	*Eating Pairs*	Capstone, 2003	
Tang, Greg	*Math-terpieces*	Scholastic Press, 2003	
Thompson, Lauren	*Little Quack*	Simon & Schuster, 2005	Activity 5-5
Wood, Audrey and Don	*The Little Mouse, the Red Ripe Strawberry, and the Big Hungry Bear*	Child's Play International, Ltd., 1993	Activity 6-16
Zaslavsky, Claudia	*Count on Your Fingers African Style*	Black Butterfly Children's Books, 2000	Project 2

Measurement and Reference Frames

Measurement

Author	Title	Publisher	Program Link
Adams, Pam	Ten Beads Tall	Child's Play International, Ltd., 1990	
Alborough, Jez	Where's My Teddy?	Candlewick Press, 2002	Activity 1-13
Allen, Pamela	Who Sank the Boat?	Paperstar, 1996	Animals All Around Theme
Barner, Bob	Bugs, Bugs, Bugs	Chronicle, 1999	
Barton, Byron	Building a House	HarperTrophy, 1990	Activity 5-12
Hightower, Susan	Twelve Snails to One Lizard: A Tale of Mischief and Measurement	Simon & Schuster, 1997	
Hoban, Tana	Is It Larger? Is It Smaller?	William Morrow & Co., 1997	
Jenkins, Steve	Actual Size	Houghton Mifflin, 2004	Animals All Around Theme
Kellogg, Steven	Much Bigger Than Martin	Dial Books for Young Readers, 1992	
Lionni, Leo	Inch by Inch	HarperTrophy, 1995	Activity 5-11
Most, Bernard	How Big Were the Dinosaurs?	Harcourt Brace, 1994	Dinosaurs Theme
Murphy, Stuart J.	A House for Birdie Mighty Maddie The Best Bug Parade	HarperCollins, 2004 HarperTrophy, 2004 HarperTrophy, 1999	 Activity 3-12 Activity 3-7
Myller, Rolf	How Big Is a Foot?	Yearling, 1991	Activity 5-7
Pinczes, Elinor J.	Inchworm and a Half	Houghton Mifflin, 2003	
Schwartz, David M.	Millions to Measure	HarperCollins, 2003	
Tompert, Ann	Just a Little Bit	Walter Lorraine Books, 1996	

Clocks and Calendars

Author	Title	Publisher	Program Link
Anno, Mitsumasa	*All in a Day*	Putnam, 1999	
Asch, Frank	*Moonbear's Bargain*	Aladdin, 2000	
Carle, Eric	*The Grouchy Ladybug*	HarperCollins Publishers, 1996	Activity 8-3
	The Very Hungry Caterpillar	Philomel Books, 1994	Calendar Routine
	Today Is Monday	Putnam Publishing Group, 1997	Calendar Routine
Carlstrom, Nancy White	*Jesse Bear, What Will You Wear?*	Aladdin, 1996	Activity 5-1
Hutchins, Pat	*Clocks and More Clocks*	Simon & Schuster, 1994	
Lobel, Anita	*One Lighthouse, One Moon*	Greenwillow Books, 2000	Seasons Theme
McCaughrean, Geraldine	*My Grandmother's Clock*	Clarion, 2002	Activity 8-2
Multiple Authors	"The Tortoise and the Hare"		Activity 6-4
Sendak, Maurice	*Chicken Soup with Rice: A Book of Months*	HarperTrophy, 1991	Calendar Routine
Schenk de Regniers, Beatrice	*May I Bring a Friend?*	Aladdin, 1989	Calendar Routine
Sweeney, Joan	*Me Counting Time*	Dragonfly, 2001	Activity 8-2
Ward, Cindy	*Cookie's Week*	Paperstar, 1997	Calendar Routine

Money

Author	Title	Publisher	Program Link
Brisson, Pat	*Benny's Pennies*	Dragonfly, 1995	Activity 6-1
Gill, Shelley and Deborah Tobola	*The Big Buck Adventure*	Charlesbridge Publishing, 2002	Activity 8-7
Hoban, Russell	*A Bargain for Frances*	HarperCollins, 1992	Activity 8-8
Hoban, Tana	*26 Letters and 99 Cents*	William Morrow & Co., 1995	
Schwartz, David M.	*If You Made a Million*	HarperTrophy, 1994	
Viorst, Judith	*Alexander, Who Used to Be Rich Last Sunday*	Atheneum, 1978	
Williams, Vera	*A Chair for My Mother*	Greenwillow, 1998	Activity 8-8
Ziefert, Harriet	*You Can't Buy a Dinosaur with a Dime*	Blue Apple Books, 2003	

Geometry

Shapes

Author	Title	Publisher	Program Link
Blackstone, Stella	*Bear in a Square*	Barefoot Books, 2000	
Burns, Marilyn	*The Greedy Triangle*	Scholastic Press, 1995	
Dodds, Dayle Ann	*The Shape of Things*	Candlewick Press, 1996	
Ehlert, Lois	*Color Zoo*	HarperCollins, 1989	Animals All Around Theme
Emberley, Ed	*Ed Emberley's Picture Pie* or *Picture Pie 2*	Little, Brown, 2006; 1996	Activity 7-4
	Go Away, Big Green Monster!	Little, Brown, 1993	
Friedman, Aileen	*A Cloak for the Dreamer*	Scholastic, 1995	
Grifalconi, Ann	*The Village of Round and Square Houses*	Little, Brown, 1986	
Hoban, Tana	*Cubes, Cones, Cylinders, & Spheres*	Greenwillow Books, 2000	
	Shapes, Shapes, Shapes	William Morrow & Co., 1986	
	So Many Circles, So Many Squares	Greenwillow Books, 1998	
Martin, Elena	*Look at Both Sides*	Capstone, 2004	
Murphy, Stuart J.	*Captain Invincible and the Space Shapes*	HarperCollins Publishers, 2001	Activity 6-3
	Circus Shapes	HarperTrophy, 1998	Activity 2-1
	Let's Fly a Kite	HarperTrophy, 2000	Activity 2-15
Pallotta, Jerry	*Icky Bugs Shapes Book*	Scholastic, 2004	
Pinczes, Elinor J.	*My Full Moon Is a Square*	Houghton Mifflin, 2002	
Tompert, Ann	*Grandfather Tang's Story*	Dragonfly Books, 1997	Activity 4-10

Spatial Relationships

Author	Title	Publisher	Program Link
Chesanow, Neil	*Where Do I Live?*	Barron's Educational Series, 1995	Project 6
Fanelli, Sara	*My Map Book*	HarperCollins, 1995	Project 6
Hennessy, B.G.	*Once Upon a Time Map Book*	Candlewick Press, 2004	Fairy Tales Theme
Hutchins, Pat	*Rosie's Walk*	Simon & Schuster, 1971	Activity 2-3
Leedy, Loreen	*Mapping Penny's World*	Henry Holt and Co., 2003	Project 6
Rabe, Tish	*There's a Map on My Lap!*	Random House Children's Books, 2002	
Sweeney, Joan	*Me on the Map*	Random House, 1998	Project 6

Patterns, Functions, and Algebra

Sorting

Author	Title	Publisher	Program Link
Algant, Paul, Jr.	How Many Snails?	⁓⁓⁓⁓⁓⁓⁓⁓⁓	
Marzollo, Jean and Walter Wick	*I Spy* book series	Cartwheel Books	Activity 5-3
Murphy, Stuart J.	*A Pair of Socks* *3 Little Firefighters*	HarperCollins, 1996 HarperCollins, 2003	Activity 4-13
Reid, Margarette S.	*The Button Box*	Penguin, 1995	Activity 1-6

Visual and Auditory Patterns

Author	Title	Publisher	Program Link
Ehlert, Lois	*Market Day: A Story Told with Folk Art*	Harcourt Books, 2000	
Harris, Trudy	*Pattern Bugs* or *Pattern Fish*	Lerner Publishing Group, 2001; 2000	Activity 5-2
Hutchins, Pat	*Changes, Changes*	Aladdin, 1987	Activity 1-2
Jonas, Ann	*The Quilt*	Puffin Books, 1994	Activity 1-2
Murphy, Stuart J.	*Beep Beep, Vroom Vroom!*	HarperCollins, 2000	Activity 1-9
Smucker, Barbara	*Selina and the Bear Paw Quilt*	Stoddart, 2002	Activity 1-2

Patterned Language

Author	Title	Publisher	Program Link
Martin, Bill Jr. and Eric Carle	*Brown Bear, Brown Bear, What Do You See?*	Henry Holt and Co., 1992	
Multiple Authors	"Chicken Licken" "I Know an Old Lady Who Swallowed a Fly" "Lazy Jack" "The Little Red Hen"		
Williams, Sue	*I Went Walking*	Houghton Mifflin Harcourt, 1996	Activity 7-15

Data and Chance

Data and Probability

Author	Title	Publisher	Program Link
LeSieg, Theodore	*Wacky Wednesday*	Random House, 1989	Activity 3-10
Murphy, Stuart J.	*The Best Vacation Ever* *Probably Pistachio*	HarperTrophy, 1997 HarperTrophy, 2001	Activity 6-5
Seuss, Dr.	*And to Think That I Saw It on Mulberry Street*	Random House, 1989	Activity 3-10

Operations and Computation

Operations

Author	Title	Publisher	Program Link
Adams, Pam	This Old Man	Child's Play International, 1991	
Hoban, Tana	*More, Fewer, Less*	HarperCollins Publishers, 1998	Activity 8-14
Jonas, Ann	*Splash!*	HarperCollins Publishers, 1997	Activity 2-14
Lewis, Kevin	*Chugga-Chugga Choo-Choo*	Hyperion, 2001	Activity 3-13
Long, Lynette	*Domino Addition*	Charlesbridge, 1996	
Merriam, Eve	*12 Ways to Get to 11*	Aladdin, 1996	Activity 7-3
Multiple Authors	"The Gingerbread Boy"		Activity 4-4
Murphy, Stuart J.	*Animals on Board* *Elevator Magic*	HarperTrophy, 1998 HarperCollins, 1997	Activity 7-6
Pinczes, Elinor J.	*A Remainder of One*	Houghton Mifflin, 1995	
Tang, Greg	*Math Fables*	Scholastic Press, 2004	Activity 7-16

Other Books Referenced in Kindergarten

Author	Title	Publisher	Program Link
Ada, Alma Flor	*Dear Peter Rabbit*	Simon & Schuster, 1997	A Working World Theme
Ahlberg, Janet and Allen	*The Jolly Postman*	Little, Brown Young Readers, 2001	A Working World Theme
Anno, Mitsumasa	*Anno's Magic Seeds*	Penguin, 1999	Project 8
Blood, Charles L. and Martin Link	*The Goat in the Rug*	Aladdin, 1990	Project 7
Brown, Marcia	*Stone Soup*	Charles Scribner, 1991	Fairy Tales Theme
Bulla, Clyde Robert	*What Makes a Shadow?*	HarperCollins, 1994	Project 8
Bunting, Eve	*Flower Garden*	Harcourt, 2000	Activity 1-8
Butterworth, Nick	*Jasper's Beanstalk*	Aladdin, 1997	Project 8
Carle, Eric	*The Secret Birthday Message*	HarperCollins, 2000	Activity 1-8
Chanin, Michael	*The Chief's Blanket*	H. J. Kramer, 1998	Project 7
Ehlert, Lois	*Planting a Rainbow*	Houghton Mifflin Harcourt, 1992	Project 8
Hall, Zoe	*Fall Leaves Fall*	Scholastic, 2000	Seasons Theme
Keats, Ezra Jack	*Pet Show!* *The Snowy Day*	Puffin, 2001 Viking Juvenile, 1962	Activity 5-13 Seasons Theme
Kindersley, Anabel and Barnabas	*Children Just Like Me: Celebrations!*	DK Publishing, 1997	Project 4
Kleven, Elisa	*Sun Bread*	Penguin Putnam, 2004	Project 4
Lankford, Mary	*Birthdays Around the World*	HarperCollins, 2002	Project 4
Morris, Ann	*Bread, Bread, Bread* *Houses and Homes*	HarperCollins, 1993 HarperCollins, 1992	Project 4 Families at Home Theme
Munsch, Robert	*Moira's Birthday*	Annick Press, 1992	Activity 1-8
Nail, James D.	*Whose Tracks Are These?*	Roberts Rinehart Publisher, 1996	Activity 5-7
Schwartz, David M.	*If You Hopped Like a Frog*	Scholastic Press, 1999	Project 2
Slobodkina, Esphyr	*Caps for Sale*	HarperTrophy, 1998	Activity 3-14

Other Books Referenced in Kindergarten

Author	Title	Publisher	Program Link
Stevens, Janet	*Tops & Bottoms*	Houghton Mifflin Harcourt, 1995	Project 8
Stevenson, Robert Louis	*A Child's Garden of Verses*	Simon & Schuster, 1999	Project 8
Sweeney, Joan	*Me and My Amazing Body*	Random House, 2000	Project 2

Teacher Resources

Author	Title	Publisher	Program Link
Braman, Arlette N.	*Kids Around the World: Cook!* and *Kids Around the World: Create!*	John Wiley & Sons, 2000; 1999	Project 4
Cook, Deanna F.	*The Kids' Multicultural Cookbook*	Ideals Publications, 2008	Project 4
Jones, Lynda	*Kids Around the World Celebrate!*	John Wiley & Sons; 1999	Project 4
Katzen, Mollie	*Pretend Soup* and *Other Real Recipes: A Cookbook for Preschoolers & Up*	Ten Speed Press, 1994; 1999	Project 4
Orlando, Louise	*The Multicultural Game Book*	Scholastic, 1993	Project 3
Wise, Debra	*Great Big Book of Children's Games*	McGraw-Hill, 2003	Project 3
Zaslavsky, Claudia	*Africa Counts* *Math Games & Activities from Around the World* and *More Math Games & Activities from Around the World*	Lawrence Hill Books, 1999 Chicago Review Press, 1998; 2003	Project 2 Project 3

Kindergarten Software List

Title of Program	Publisher	Contact Information	Format	Grade Level	Description
The Graph Club™ 2.0	Tom Snyder	80 Coolidge Hill Road Watertown, MA 02427-5003 1-800-342-0236 www.tomsnyder.com	Macintosh/Windows CD-ROM	K–4	Children conduct surveys in school, graph the results, and print the graph for display in the classroom.
Graphers	Sunburst (Hyperstudio)	Sunburst Technology 1550 Executive Drive Elgin, IL 60123 1-888-492-8817 www.sunburst.com	Macintosh/Windows CD-ROM	K–4	Children collect, organize, sort, and describe data, and construct and work with various types of graphs.
I Spy: School Days	Scholastic	Scholastic Inc. 555 Broadway New York, NY 10012-3999 www.scholastic.com	Macintosh/Windows CD-ROM	K–4	Children follow a Treasure Map to build and strengthen problem-solving skills.
I Spy: Spooky Mansion	Scholastic	See above.	Macintosh/Windows CD-ROM	K–4	Children explore a Spooky Mansion to build and strengthen problem-solving skills.
I Spy: Treasure Hunt	Scholastic	See above.	Macintosh/Windows CD-ROM	K–4	Children solve riddles to build and strengthen problem-solving skills.
JumpStart Math for Kindergarteners	Knowledge Adventure	2377 Crenshaw Blvd., Suite 302 Torrance, CA 90501 (310) 533-3400 www.knowledgeadventure.com	Macintosh/Windows CD-ROM	K	Children count, add, subtract, make equations, make patterns, sort objects, and solve problems.
Math Blaster® Ages 5–7	Knowledge Adventures	See above.	Macintosh/Windows CD-ROM	K–2	Children count and recognize money, tell time, and solve number stories.
Math Missions: Grades K–5	Tom Snyder	See above.	Macintosh/Windows CD-ROM	K–5	Children use money to help run businesses and play games. Math concepts covered include: counting and place value; measurement, time, and money; sorting and classifying; addition and subtraction; and shapes.

Title of Program	Publisher	Contact Information	Format	Grade Level	Description
Mighty Math™: Carnival Countdown	Edmark	P.O. Box 97021 Redmond, WA 98073-9721 1-800-691-2985 www.riverdeep.net/edmark	Macintosh/Windows CD-ROM	K–2	Children practice problem solving, place value, addition and subtraction, logic, sorting and classification, pattern recognition equivalencies, multiplication and division, and 2-D geometry in a carnival adventure.
Mighty Math™: Zoo Zillions	Edmark	See above.	Macintosh/Windows CD-ROM	K–2	Children solve story problems, count money, make change, add and subtract facts, and explore early 3-D geometry.
Millie's Math House	Edmark	See above.	Macintosh/Windows CD-ROM	Pre-K–2	Children explore numbers, shapes, sizes, patterns, and addition and subtraction.
Piggy in Numberland	Learning in Motion, Inc.	500 Seabright Avenue, Suite 105 Santa Cruz, CA 95062 1-800-560-5670 www.learn.motion.com	Macintosh/Windows CD-ROM	Pre-K–1	Children practice and explore number sense and counting skills, addition and subtraction, numerical reasoning, and geometric shapes in Piggy's world.
Piggy's Birthday Present	Learning in Motion, Inc.	See above.	Macintosh/Windows CD-ROM	K–2	Children practice skills in number recognition, number sequences, comparisons of numbers, and simple addition and subtraction.
Thinkin' Things Collection 1	Edmark	Riverdeep, Inc. 100 Pine Street, Suite 1900 San Francisco, CA 94111 1-888-242-6747 Email: info@riverdeep.net www.edmark.com	Macintosh/Windows CD-ROM	Pre-K–3	Children explore activities and games that focus on memory development, logic, visual and spatial thinking, musical memory, and problem solving.
To Market, To Market	Learning in Motion, Inc.	See above.	Macintosh/Windows CD-ROM	Pre-K–3	Children explore and compare sets of objects and use addition and subtraction to make the sets equal to, more than, or less than one another.

The *Early Childhood EM Games* are available on a Family Version CD or in an online version so children can play the same games in the Kindergarten classroom and at home with their families. Some of the games on the CD and online version include:

△ *Paper Money Exchange*
△ *Top-It*
△ *Spin a Number*
△ *Disappearing Train Game*
△ *Monster Squeeze*
△ *Plus or Minus Game*

Games for the Classroom

The activities in the *Teacher's Guide to Activities* feature numerous *Kindergarten Everyday Mathematics* games that are made in the classroom. (See the K–2 Games Correlation Chart on pages 96 and 97 of this book for a complete list.) This section lists additional games that you might obtain for your classroom to reinforce mathematics concepts and skills.

Most of these games are categorized by the mathematics skill that children are most likely to use when playing them. However, many of the games involve a variety of skills. Most games, especially those that use spinners or dice, also incorporate probability and chance concepts.

This section also includes lists of cooperative games, card games, and other games that involve mathematics.

Counting and Number Recognition

Game and Publisher	Skills	Description
Buggo™ (Ravensburger®)	Counting; comparing numbers	Players search for the most beetles but must start over if they turn over a buggo.
Bzz Out™ (Gamewright®)	Comparing numbers; counting	The player with the highest card wins the round and gets a honey pot.
Chutes and Ladders™ (Milton Bradley®)	Counting and reading numbers to 100	Players try to be the first to climb to 100.
Cootie™ (Hasbro®)	Counting; using probability	Players toss a die to get the parts to build a Cootie.
4-Way Count Down!™ (Cadaco®) or Shut the Box (Square Roots)	Recognizing numbers; finding number equivalents	Players try to be the first to turn down all their numbers.
Fowl Play™ (Gamewright®)	Counting; using number sense	Players turn over cards to get chickens, but if they get a wolf they lose everything.
Hi, Ho! Cherry-O™ (Parker Brothers®)	Counting 1–10; adding; subtracting	Players try to be the first to pick all their cherries.
Rat-a-Tat Cat (Gamewright®)	Comparing numbers; adding; using strategy	Players try to get the fewest number of rats in their collection of cards.
Sorry!™ (Hasbro)	Counting; recognizing numbers; using strategy	Players move their pawns around a gameboard and safely get back to "home."
Spit!™ (Patch Products®)	Making number sequences	Players try to be the first to use all their cards.
Trouble™ (Milton Bradley)	Counting; using strategy	Players try to be the first to get all their pegs home.

Money

Game and Publisher	Skills	Description
Monopoly, Jr.™ (Parker Brothers)	Counting; using money; making change	Players move around the board setting up booths and collecting money.

Attributes, Patterns, Geometry

Game and Publisher	Skills	Description
Big Top™ (Gamewright®)	Finding missing attributes	Players find the missing animal and color.
Candy Land™ (Milton Bradley)	Matching colors	Players try to be the first to arrive at the castle.
Dog Dice® (Gamewright®)	Coordinating two attributes; matching attributes	Players try to fill up the Bingo-type board.
Guess Who?™ (Milton Bradley)	Using attribute clues	Players determine the secret person by eliminating categories.
Hisss™ (Gamewright®)	Matching; collecting; counting	Players make snakes to collect the most cards.
I Spy Memory Game™ (Briarpatch®)	Matching attributes; making visual discriminations	Players find a match for an object that may have a different size, position, or context.
Kitty Corners™ (Gamewright®)	Coordinating two attributes; matching attributes	Players try to cover their boards.
Mystery Garden (Ravensburger)	Recognizing common attributes; counting	Players collect three secret mystery tiles.
Peanut Butter and Jelly™ (Fundex®)	Matching; sequencing; recognizing numbers	Players try to be the first to make a sandwich.
Slamwich® (Gamewright®)	Matching; patterning	Players collect cards by making "slamwiches" and double deckers.
Uno® (Mattel®)	Matching numbers; recognizing colors; using strategy	Players try to get rid of all their cards. (There is no need to keep score with young children.)

Strategy

Game and Publisher	Skills	Description
Battleship™ (Milton Bradley)	Using coordinates; strategy	Players find and sink their opponent's fleet.
Clue, Jr.™ (Parker Brothers)	Counting; using strategy	Players find the hidden object by moving around the board and collecting clues.
Connect Four™ (Milton Bradley)	Using strategy; spatial reasoning	Players try to get four in a row.

Game and Publisher	Skills	Description
Mastermind for Kids® (Pressman Toy®)	Using strategy	Players try to figure out the hidden animals.
Junior Labyrinth (Ravensburger)	Using strategy; spatial relationships	Players try to travel the maze to find the treasure and avoid getting trapped.
Yahtzee, Jr.™ (Hasbro)	Matching; making sets; using strategy and probability	Players try to fill the most spaces on the board.

Cooperative Games

Children work together to solve a problem or reach the goal; there are no winners or losers.

Game and Publisher	Skills	Description
Granny's House® (Family Pastimes®)	Counting; planning; using strategy	Players try to land on the lucky spots to get to Granny's house without trouble.
Max® (Family Pastimes)	Cooperating to make decisions; using strategy; counting	Players try to keep Max from catching the little creatures.
Princess® (Family Pastimes)	Thinking flexibly and creatively; counting	Players try to save the Princess from the spell.
Round-Up® (Family Pastimes)	Planning cooperatively; using strategy; counting	Players round up the horses and bring them back to the ranch.
The Secret Door® (Family Pastimes)	Using time; matching cards; using strategy	Players try to find the missing valuables before midnight.
Sleeping Grump® (Family Pastimes)	Counting; using strategy	Players try to get the treasure before the Grump wakes up.

Other Games (available from many sources)

Game	Skills	Description
Bingo	Reading 2-digit numerals	Players try to cover 5 in a row.
Checkers	Using strategy	Players try to capture all of their opponent's checkers.
Chinese Checkers	Using strategy	Players try to be the first to move all of their marbles to the opposite side.
Dominoes*	Counting; using strategy	Players try to find matching numbers for all of their domino tiles.
Mancala	Counting; using strategy	Players try to capture the most stone counters.

*NOTE: Many games can be played with domino tiles. Muggins and Mexican Train are two familiar games. (Many Internet sites give rules for domino games.)

Card Games

Search the Internet for "children's card games." There are many sites that provide rules for card games. Many card games can be played with a regular deck of cards; games are also available in decks unique for each game.

Crazy Eights
Skills: Matching; recognizing numbers
Description: Players try to get rid of all their cards.

Go Fish
Skill: Making matches and pairs
Description: Players try to make the most pairs.

Old Maid
Skill: Making matches
Description: Players avoid getting stuck with the "Old Maid" card.

Slap Jack
Skill: Making matches
Description: Players capture matches.

Spit
Skill: Sequencing numbers
Description: Players try to get rid of their cards by sequencing numbers.

War*
Skill: Comparing numbers
Description: Players try to capture the most cards.

*The *Everyday Mathematics* version of this game is *Top-It*.

Everyday Mathematics Take-Home Games

The *Everyday Mathematics Early Childhood Family Games Kit* is a take-home collection of *Everyday Mathematics* games that families can play together to develop basic skills, critical thinking, and problem solving. The kit includes gameboards, game pieces, and directions for games that include the following:

▷ *Spin a Number*
▷ *Disappearing Train Game*

▷ *Monster Squeeze*
▷ *Plus or Minus Game*

A *Teacher's Guide to Games* is also available.

Puzzles

Puzzles develop children's understanding of spatial relationships and problem-solving skills. All classrooms should have a variety of puzzles, including jigsaw puzzles, floor puzzles, and some of the following:

Polydron™ (Polydron Limited) Geometric shapes for making 2- and 3-dimensional designs and patterns.

Tangrams A set of 7 shapes that can be arranged to make different pictures.
Tangoes (Rex Games) A game based on Tangrams.

Ideas for Newsletters

These short descriptions can be used as part of your ongoing communication with families. Use them as is or personalize them for your classroom newsletter. Choose the descriptions that appeal to you and apply to your classroom. Frequent, informal communication with families reinforces the idea that mathematics is everywhere and is a part of our everyday lives!

The first four suggestions can be used at any time during the school year. The others are based on the activity sequence in the *Teacher's Guide to Activities*. In addition to these newsletter ideas, there are four full-page blackline master family letters on pages 74–78 of this book for you to use at different points in the year.

Photographs can be useful for communication in newsletters, too. You might take pictures of children at work during math time or in the Math Center and include one or more of these photos in your newsletter along with captions explaining what children are doing and why they are doing it.

Serving Food

Children enjoy helping with snack preparations at school. At home, your child can help set the table by placing one plate, cup, and set of utensils at each place and then counting the total number of items on the table. Also work with your child to figure out how to distribute the same amount of food to each family member.

Use any time.

Preparing Food

The children in our class enjoyed our recent cooking project. You can use mathematics as you cook together at home, too. Decide what to cook, then help your child read the recipe, identify the measuring tools needed, and measure the appropriate amounts. Your child can also set a timer or look at a clock to monitor the cooking time.

Use after you have done a cooking project in class.

Songs and Rhymes

(You might want to include the words to favorite rhymes or counting songs along with a note such as this:)

Our class loves to sing and act out rhymes and fingerplays such as "One, Two, Buckle My Shoe."

Rhymes like this one help children learn number words and counting sequence, as well as concepts related to position and order. We also sing lots of counting songs. Encourage your child to teach family members some of the rhymes and songs he or she has learned. You will find many books of songs and rhymes in your library!

Use any time. Vary to match a particular song or rhyme.

Sending Games Home

(It is a good idea to briefly explain the mathematical value of the game, in addition to the game directions. You might also include a sheet for families to share their comments about the game.)

Games are an important part of *Everyday Mathematics*. In school, we play games frequently to reinforce skills and concepts and to develop problem-solving strategies in a fun way. Children enjoy teaching family members the games we play at school.

Included in this bag are all of the materials you need to play [*name of game*]. As children play this game, they practice [*mathematical skill*] and develop their understanding of [*mathematical concept*]. Play the game at home with your child. Then place all of the materials back in this bag and return the bag to school by [*date*]. Use the comment sheet to share how you liked the game and any other comments you might have about it.

Use this note to accompany games that you send home.

Introduction to Money

Working with coins is an important part of our mathematics program. To build children's familiarity with money, let your child borrow the coins from your purse or pocket and sort them by attributes—color, size, markings, and so on. Also let your child help you pay for items at the store. (Children love to receive change!) Do not be concerned with coin values at this time. Children will learn more about coin values and exchanges later in Kindergarten and in first grade.

Use near the end of Section 1.

Telling and Solving Number Stories

A "number story" is another name for a "story problem." We have been creating and solving many number stories at school. Encourage your child to figure out answers to real-life situations such as "You and your sister can each have one cookie. How many cookies should I take out of the cookie jar?" or "I already put 2 napkins on the table. How many more do I need for our family?" Your child will enjoy making up number stories for you to solve.

Number stories are introduced in Activity 2-14. You might use this note a few weeks later, after your class has had many experiences with them.

Number Books

Children recently finished making their own number books. They worked very hard at deciding what to draw and then drawing the correct number of pictures to match the number on each page. They also practiced forming numerals, although this remains difficult for many children. Children will have many more opportunities to practice number writing over the course of the year as their fine-motor skills develop.

Use when children's Number Books, which are introduced in Activity 3-1, are complete.

Games

Our class has enjoyed playing some new math games recently. (You might ask your child about *Spin a Number, Teen Partner,* and *Monster Squeeze.*) Children will continue playing these games and others throughout the year. Games reinforce skills and concepts and are an important part of *Everyday Mathematics.*

Use near the end of Section 3.

Estimating

Making estimates about the items in our Estimation Jar is a popular class activity, and children are gradually getting better at making reasonable estimates, rather than wild guesses. We often use a "reference jar" (with a known quantity of the same objects) to help us estimate. When we estimate at school, we emphasize that the goal is to make a "smart guess," not to get the exact number. Good estimation skills are related to good number sense.

Use after children have had several opportunities to estimate, perhaps sometime during Section 4 or Section 5.

Charitable Collections

(Many schools organize charitable collections. These experiences provide rich opportunities for children to count, sort, classify, and collect data. Modify the samples below to fit the specifics of your situation.)

Food Drive

The children are very proud of the number of cans of food we have collected for the school food drive, thanks to all of you! We have been counting all of these cans—[number of cans] so far—and recording the numbers on a grid. We record the number at the end of each count, and then count on from the last recorded number every time we take stock of the collection. Each time we count, we update our recording grid. We have also arranged the cans in rows of 10 and counted by 10s. Now we are weighing the cans on a scale (several at a time) and using a calculator to find the total number of pounds the cans weigh.

Mitten Collection

We have collected [number of pairs] pairs of mittens for the school mitten tree. The children counted the pairs by 2s, then counted the individual mittens by 1s. We discovered that we had [number of mittens] mittens, no matter how we counted! Your child might enjoy counting the socks in his or her sock drawer in a similar way.

Use in accordance with any charitable collection effort.

Graphing Sums of Dice Throws

The latest craze in our classroom is graphing dice throws. You may have noticed a completed dice-throw grid in your child's backpack. To do this activity, children roll two dice, figure out the total, and mark the appropriate space on their dice-throw grids. Children continue until one number "wins" (one column is filled). We recorded each child's winning number on a class grid and talked about why 6, 7, and 8 did so much better than the other numbers. It didn't take children long to figure out that they can't get a sum of 1 when throwing two cubes that have at least one dot on each! During this activity, children were practicing addition and exploring probability!

Use after Activity 4-8.

More Number Stories

As they continue to tell and solve number stories, children are beginning to learn strategies for solving addition and subtraction problems and are becoming acquainted with mathematical language and symbols. Number stories provide a bridge from spoken language to mathematical  symbolic language. Children are welcome to share with the class number stories that they create at home. They might illustrate their stories or share them orally. These number stories can provide a mathematical "show and tell" experience for the class!

Use anytime after Activity 4-15. Use more than once, if desired.

Calculators

Most children love to use calculators, which makes them feel very grown-up. Calculators also make it possible for young children to display and read numbers before they are skilled at writing them. At school we have been building children's familiarity with calculators by using calculators to display the answers to questions such as "How many feet does a cat have?" Children are also exploring the various symbol keys on the calculators (especially the +, −, and = keys) and are learning to use calculators to count on and back from a given number.

Use after Activity 5-5.

Graphing

Collecting data and using it to create and interpret graphs provides many opportunities for children to count and compare quantities. This week we made a Pet Bar Graph at school. It was interesting to see whether children have pets, what kinds of pets they have, and to figure out which pets were represented the most and least on our graph. We have used real-life data from the class to create lots of graphs at school. Children's own class surveys are the source of many graphing activities.

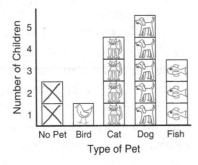

Use after Activity 5-13.

Coins and Values

We have been learning about coin names and values at school. So far, we've talked about pennies, nickels, and dimes. We will discuss quarters soon. Children enjoy figuring out how many pennies are in a nickel or a dime, how many nickels are in a dime, and how many dimes make a dollar, but they still need reminders and practice. They will continue to develop these skills throughout this year and in first grade.

Use toward the end of Section 6.

Following Up on the 100th Day of School

(You might involve children in composing a 100th Day report to families. See the sample below.)

As you know, last week we celebrated our 100th Day of learning together in Kindergarten. Our activities included the following:

- Sharing our "100 objects" collections
- Reading stories about the number 100
- Counting by 1s, 5s, and 10s to 100
- Bundling sticks in 10 groups of 10 and creating one bundle of 100
- Writing numbers on a 100 grid
- Mixing 100 snack items and eating our Hundred Day Trail Mix
- Figuring out different ways to group and count 100 items.

Use after the 100th Day of school.

Class Collection

(You might accompany a note about your class collection with a picture of the final collection.)

Thanks to everyone's contributions, we have collected a total of 356 milk bottle tops! To track our growing total, we counted how many caps were in the collection basket several times each week. We kept a record by counting on from the previous total and recording it on a chart. It has been fun to monitor the growing total in different ways. It has also been good practice to make groups of 10 and count those groups by 10s. When we began our collection we predicted that we would collect more green caps, but on the last collection day we sorted the caps by color and discovered that we had more red caps than any other color!

Use upon completion of a class collection. The class begins a collection in Activity 7-2.

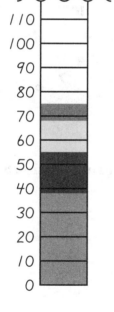

Number Scrolls

Children have begun to make number scrolls by writing numbers on a blank number grid. This provides practice with writing numbers and helps children develop a sense of number patterns and place value. Children love watching their scrolls grow!

Use in conjunction with children's number scrolls, which are introduced in Activity 7-10.

Weaving

We have been weaving as part of our mathematics curriculum! In addition to being an enjoyable and satisfying art activity, the process of weaving provides very concrete patterning experiences, develops spatial orientation skills, and reinforces orientation words such as *over/under, in/out,* and *back/forth.* It also teaches children to be patient and to persevere.

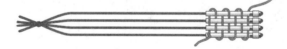

Use after Project 7.

Name Collections

At school we are talking about the idea that numbers can be shown in different ways, which we call *equivalent names for numbers.* For example, 6 can be shown as six pennies; the word *six;* six tally marks; or a number sentence like 1 + 5 or 2 + 4. We refer to a group of equivalent names for numbers as a *Name Collection.* Thinking about numbers in this way helps children develop solid number sense.

Today your child is bringing home a bead string—a tool we used at school to explore the concept of equivalent names for numbers. As your child moves the beads to create various groupings (for example, on a string with 5 beads: 1 and 4; 2 and 3; 0 and 5), he or she is reinforcing the idea of equivalent names for numbers. Encourage your child to describe the number combinations he or she makes with the bead strings.

Send home with children's bead strings after Bead String Name Collections, Activity 7-16.

Clocks

Children require time and practice in order to tell time on an analog clock (a clock with hands). In Kindergarten, we begin by focusing on the hour hand. You may be surprised at how accurately you can tell time using just the hour hand! Involve your child as you check the time, emphasizing the position of the hour hand. For example, notice when the hour hand is pointing directly to 4 and mention that it is 4 o'clock. Similarly, when the hour hand is slightly before or after the 4, explain that it is "just before" or "a little after" 4 o'clock. Children will learn to use the minute hand to tell time as part of first and second grade activities.

Use during Section 8.

Family Letter Masters

Contents

Dear Families,

Welcome to *Kindergarten Everyday Mathematics*, a program created by the University of Chicago School Mathematics Project. This program is based on research and experience that shows that young children are capable of far more mathematics learning in Kindergarten than was previously believed, provided that the content is presented and explored in age-appropriate ways.

Over the course of the year, your child will do many hands-on activities related to a range of mathematical topics, including counting, numeration, measurement, geometry, patterns, sorting, data collecting, and calculator use. Classroom routines such as keeping track of the days of school, helping with attendance, and observing and graphing weather and temperature give children real-life opportunities to develop and refine mathematics skills and become "math thinkers." Periodically, you will receive "Home Links" which suggest ways to help your child by doing mathematics activities at home.

The playful mathematics activities that make up *Kindergarten Everyday Mathematics* are meaningful and productive and are designed to help children build a solid understanding of mathematical skills and concepts. Research has shown that children have more success with written and symbolic mathematics in later grades if they have a Kindergarten experience that builds a strong foundation based on experience and understanding.

Everyday Mathematics is a Kindergarten through Grade 6 curriculum. Content in the early grades begins with concrete experiences. Topics, concepts, and skills are revisited in varied ways and contexts over time, integrating new learning with previous knowledge and experiences. Children will revisit and build upon skills and concepts throughout the Kindergarten year. They will continue to develop their understanding of topics that they encounter in Kindergarten as they move through later grades.

As children participate in *Kindergarten Everyday Mathematics* activities, they will find that mathematics is useful, enjoyable, varied, and meaningful. Just as telling stories and reading books to children helps foster a love of reading, your involvement in your child's ongoing mathematics experiences will help him or her develop lasting excitement, confidence, and competence in math!

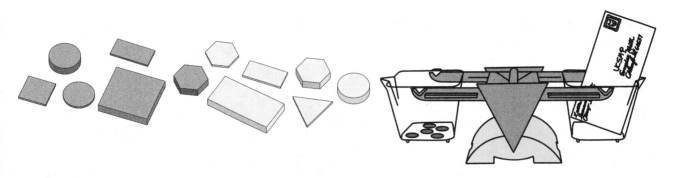

Introducing Our Classroom Routines

Dear Families,

Routines are an important part of daily life in our classroom. They provide children with security and predictability, help build classroom community and collaboration, and make aspects of classroom life (such as attendance and classroom jobs) run more smoothly. They also provide valuable and meaningful opportunities to integrate mathematics and other subject areas into everyday activities. Children are involved in a variety of mathematical experiences as they carry out the following classroom routines:

▷ The **Number of the Day Routine** develops counting and other numeration skills, such as number writing and place value. Beginning with the first day of school, we add a new number each day to our Growing Number Line. As the line grows, children become increasingly aware of number patterns and begin to use the Number Line as a tool for solving problems and playing mathematics games.

▷ The **Attendance Routine** provides a meaningful way to count and work with numbers and data as children check in every day and count the number of children who are present and absent.

▷ The **Job Chart Routine** gives children responsibility for various classroom tasks, which builds confidence and pride and provides practice with a variety of different skills. As children begin to understand the pattern of rotation on the job chart, they can predict when they will have a particular job.

▷ The **Monthly Calendar Routine** helps children develop a sense of time as they keep track of the day of the week, date, and month.

▷ The **Daily Schedule Routine** helps children gain familiarity with time periods during the day as they order and track their day at school.

▷ The **Weather and Temperature Routines** allow children to record and track temperature measurements and weather observations. These routines provide rich opportunities for graphing and interpreting data and help children develop awareness about seasonal changes and tools used to measure weather.

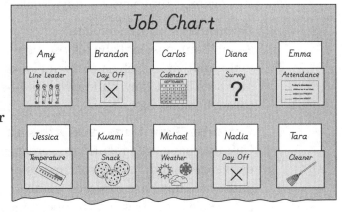

▷ In the **Survey Routine**, children respond and record their answers to a "question of the day" (or week). This provides regular experiences with collecting, graphing, and interpreting data.

When you talk to your child about the school day, you might begin by asking what job he or she had, how many days he or she has been in school, or ask about the survey question and results. Questions such as these often encourage children to share more information than "What did you do at school today?"

Mid-Year Update

Dear Families,

As you have probably noticed from the Home Links, a variety of topics are included in the *Kindergarten Everyday Mathematics* program. During the past few months, our class has begun a journey through six mathematical strands with activities such as the following:

Number and Numeration Children count every day in different ways and from different numbers—by 1s; forward and backward; and by 5s and 10s. Children have had lots of practice with reading and writing numerals and with comparing numbers through daily routines, games, and other activities.

Operations and Computation Children explore addition and subtraction through concrete activities, games, and number stories. They are encouraged to develop multiple strategies for solving addition and subtraction problems. Counting using fingers or other objects is still very acceptable and useful!

Data and Chance Children collect, organize, and display classroom data through the daily Weather, Temperature, and Survey Routines. They also work with data and graphing in activities such as graphing dice rolls. We have talked a bit about probability by describing the likelihood of events using formal and informal language such as *certain, definite, possible, impossible,* and *unlikely*.

Measurement and Reference Frames Children began with direct comparisons of the size of two objects and then moved to using manipulatives and their own feet to measure the length of various items. These experiences with nonstandard measurement units and tools lay groundwork for understanding the need for standard units of measure, which children are beginning to use. In addition, children have been learning about coins and their values, time periods (day, week, and month), and temperature measures.

Geometry Children explore 2- and 3-dimensional shapes through manipulatives—such as pattern blocks, attribute blocks, and building blocks—and through games like *I Spy*.

Patterns, Functions, and Algebra Children have been identifying, creating, and extending sound, movement, and color patterns. They have explored number patterns on our Growing Number Line and Class Number Grid. Children have also explored sorting objects according to a variety of attributes.

This list is just a small sampling! Every day children are engaged in a variety of experiences that are fun and meaningful and are designed to develop skills in a broad range of mathematical topics.

End-of-Year Message

Dear Families,

It is so exciting to look back to the beginning of the year and see how much children have grown in their understanding of mathematics! The *Everyday Mathematics* curriculum we used this year is based on the idea that we encounter and use mathematics in our everyday lives.

So, over the summer be on the lookout for chances to "talk and think" math with your child in the context of their everyday experiences. This list of Summer Math Ideas may help you to incorporate mathematics into your summer activities in fun and interesting ways. You may want to post the list on your refrigerator for handy reference.

Have a wonderful, math-filled summer!

Summer Math Ideas

▷ Continue to use the Home Links and *Mathematics at Home* books as resources for ideas to "do math" together.

▷ Play card or board games, including *Everyday Mathematics* games sent home during the year.

▷ Choose some math-related books when you go to the library.

▷ Keep track of your activities on the calendar. Count how many more days there are until a family vacation or special event.

▷ Measure and record how tall your child is at the beginning of the summer, then again at the end of summer. Compare the measures to see how much he or she has grown.

▷ Visit a local zoo or farm and observe and describe the many different kinds and sizes of animals, as well as the patterns of animal markings.

▷ Go berry picking and count and compare numbers of berries. Weigh the berries and compare the weights of different containers.

▷ Cook together. Cooking involves reading numbers, counting, measuring ingredients, measuring time, and sequencing.

▷ Prepare a picnic with various geometric shapes (sandwiches in triangular, rectangular, and square shapes, cylindrical cans, cubes of cheese, and so on).

Summer Math Ideas

▷ Set up an obstacle course at the park or in your yard. Go *around* bushes, *under* lawn chairs, *over* the toy blocks, and so on. Draw a map of the course.

▷ Read and follow the maps you find at the zoo, museum, or shopping mall.

▷ Have your child help you look up, copy, and dial phone numbers when arranging play dates or planning an outing.

▷ Allow your child to pay and receive change at the store.

▷ Look for numbers all around—on the mailbox, telephone, book pages, houses, restaurants, gas pumps, and so on.

▷ Incorporate mathematics on a family trip: How many miles will you travel? How many days will you be gone? How much money do you need for the gas or lodging?

▷ When traveling on a car trip, watch for road signs and help your child tally the shapes seen. You might also look for numbers in order as you read the signs.

▷ Measure objects using non-standard units (feet, hands, blocks, pencils), then compare the findings. Children might also enjoy using rulers and measuring tapes.

▷ Look for geometric shapes around the house, at the grocery store, at the park, in the mall, or anywhere you go! Notice both 2- and 3-dimensional shapes.

▷ Write numbers in the sandbox, dirt, play dough, or on the sidewalk with chalk. Use play dough to create 3-dimensional numbers!

▷ Use a thermometer or other source to find the temperature on different days.

▷ Create and solve number stories with family members.

Observe and describe the patterns of animal markings.

Theme Masters

Contents

Dinosaur Teeth

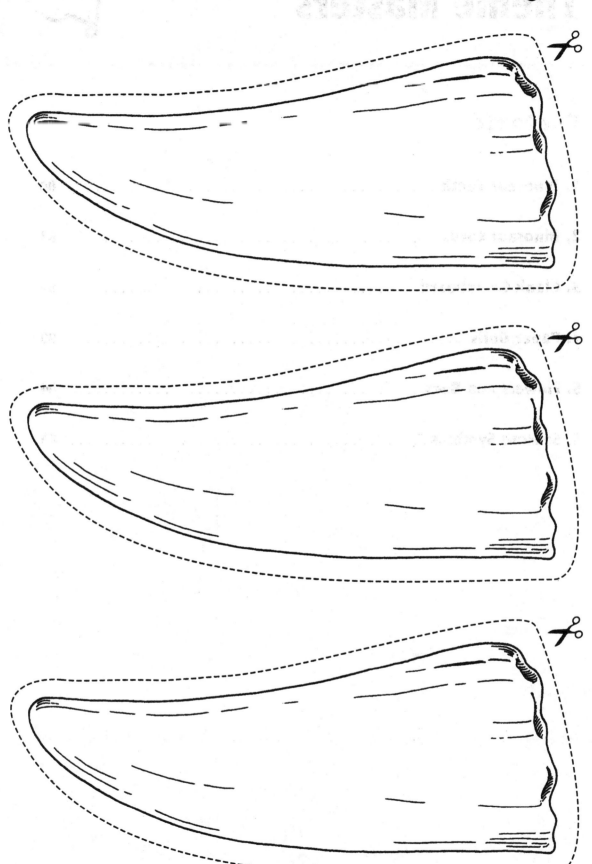

Use with Dinosaurs Theme.

Dinosaur Cards

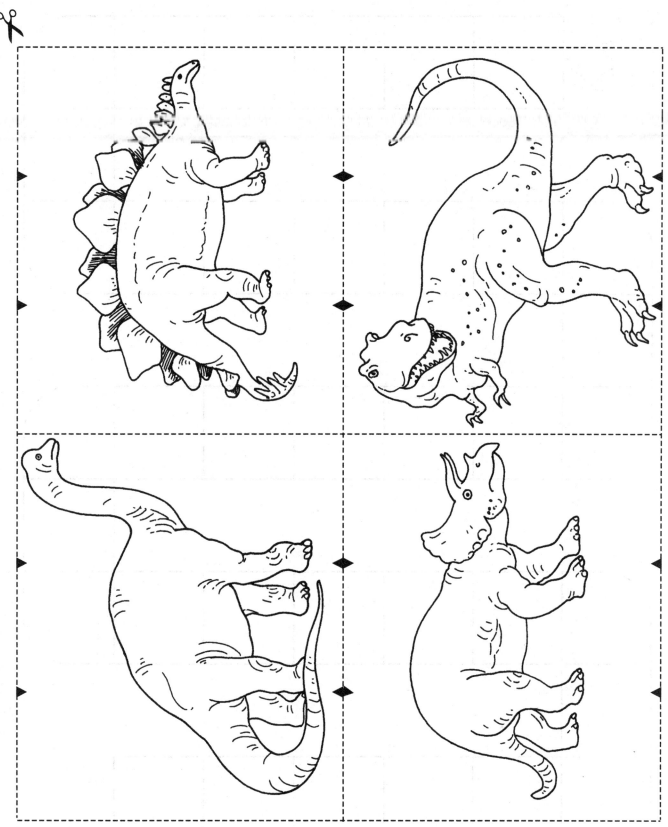

Blank Gameboard

Start

Use with Dinosaurs, Fairy Tales, and Growing Things Themes.

Paper Dolls

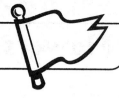

Use with Fairy Tales and Families at Home Themes.   83

Use with Growing Things Theme.

Seasons Symbols

Glossary

A

addend Any one of a set of numbers that are added. For example, in 5 + 3 + 1, the addends are 5, 3, and 1.

analog clock (1) A clock that shows the time by the positions of the hour and minute hands. (2) Any device that shows time passing in a continuous manner, such as a sundial. Compare to *digital clock*.

Analog clock

array (1) An arrangement of objects in a regular *pattern*, usually rows and columns. (2) A rectangular array. In *Everyday Mathematics,* an array is a rectangular array unless specified otherwise.

attribute A feature of an object or common feature of a set of objects. Example of attributes include size, shape, color, and number of sides. Same as property.

attribute blocks A set of blocks in which each block has one each of four *attributes* including color, size, thickness, and shape. The blocks are used for attribute identification and sorting activities. Compare to *pattern blocks*.

B

bank draft A written order for the exchange of money. For example, $1,000 bills are no longer printed so $1,000 bank drafts are issued. People can exchange $1,000 bank drafts for smaller bills, perhaps ten $100 bills.

bar graph

bar graph A graph that uses horizontal or vertical bars to represent data.

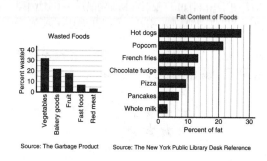

Bar graphs

C

calendar (1) A *reference frame* to keep track of the passage of time. Many different calendars exist, including the Gregorian calendar currently used by most of the Western world, the Hebrew calendar, the Islamic calendar, and others. (2) A practical model of the reference frame, such as the large reusable Class Calendar in *Kindergarten Everyday Mathematics.* (3) A schedule or listing of events.

cent A penny; $\frac{1}{100}$ of a dollar. From the Latin word *centesimus,* which means a hundredth part.

centimeter (cm) A metric unit of length equivalent to 10 millimeters, $\frac{1}{10}$ of a decimeter, and $\frac{1}{100}$ of a meter.

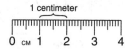

chance The possibility of an outcome occurring in an uncertain event. For example, in tossing a coin there is an equal chance of getting heads or tails.

circle The set of all points in a plane that are equally distant from a fixed point in

Disk

the plane called the center of the circle. The distance from the center to the circle is the radius of the circle. The diameter of a circle is twice its radius. Points inside a circle are not part of the circle. A circle together with its interior is called a disk or a circular region.

column A vertical arrangement of objects or numbers in an *array* or a table.

column

cone A *geometric solid* with a circular base, a vertex *(apex)* not in the plane of the base, and all of the line segments with one endpoint at the apex and the other endpoint on the circumference of the base.

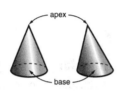

apex

base

Cones

counting numbers The numbers used to count things. The set of counting numbers is {1, 2, 3, 4, ...}. Sometimes 0 is included, but not in *Everyday Mathematics*. Counting numbers are in the sets of whole numbers, integers, rational numbers, and real numbers, but each of these sets includes numbers that are not counting numbers.

cube (1) A regular polyhedron with 6 square faces. A cube has 8 vertices and 12 edges.

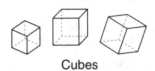

Cubes

(2) In *Everyday Mathematics,* the smaller cube of the base-10 blocks, measuring 1 cm on each edge.

cup (C) A U.S. customary unit of *volume* or capacity equal to 8 fluid ounces or $\frac{1}{2}$ pint.

cylinder A *geometric solid* with two congruent, parallel circular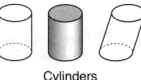

Cylinders

regions for bases and a curved face formed by all the segments with an endpoint on each circle that are parallel to a segment with endpoints at the centers of the circles. Also called a circular cylinder.

D

data Information that is gathered by counting, measuring, questioning, or observing. Strictly, data is the plural of *datum,* but data is often used as a singular word.

decade Ten years.

degree Celsius (°C) The unit interval on Celsius thermometers and a metric unit for comparing *temperatures*. Pure water at sea level freezes at 0°C and boils at 100°C.

degree Fahrenheit (°F) The unit interval on Fahrenheit thermometers and a U.S. customary unit for comparing *temperatures*. Pure water at sea level freezes at 32°F and boils at 212°F. A saturated salt solution freezes at 0°F.

difference The result of subtracting one number from another. For example, the difference of 12 and 5 is $12 - 5 = 7$.

digit (1) Any one of the symbols 0, 1, 2, 3, 4, 5, 6, 7, 8, and 9 in the base-ten numeration system. For example, the numeral 145 is made up of the digits 1, 4, and 5. (2) Any one of the symbols in any number system. For example, A, B, C, D, E, and F are digits along with 0 through 9 in the base-16 notation used in some computer programming.

digital clock A clock that shows the time with numbers of hours and minutes, usually separated by a colon. Compare to *analog clock*.

Digital clock

dollar The basic unit in the U.S. monetary system, equal to 100 *cents*.

E

equivalent Equal in value but possibly in a different form. For example, $\frac{1}{2}$, 0.5, and 50% are all equivalent.

equivalent names Different ways of naming the same number. For example; 2 + 6, 4 + 4, 12 − 4, 18 − 10, 100 − 92, 5 + 1 + 2, eight, VIII, and ⧾⧾⧾ /// are all equivalent names for 8.

estimate (1) An answer close to, or approximating, an exact answer. (2) To make an estimate.

F

foot (ft) A U.S. customary unit of length equivalent to 12 inches or $\frac{1}{3}$ of a yard.

fraction A number in the form $\frac{a}{b}$ or a/b, where a and b are whole numbers and b is not 0. A fraction may be used to name part of an object or part of a collection of objects; to compare two quantities; or to represent division. For example, $\frac{12}{6}$ might mean 12 eggs divided into 6 groups of 2 eggs each; a ratio of 12 to 6; or 12 divided by 6.

fulcrum (1) The point on a mobile at which a rod is suspended.

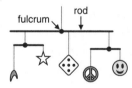

(2) The point or place around which a lever pivots.

(3) The center support of a pan balance.

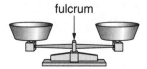

function A set of ordered pairs (x, y) in which each value of x is paired with exactly one value of y. A function is typically represented in a table, by points on a coordinate graph, or by a rule such as an equation. For example, for a function with a rule of "double," 1 is paired with 2, 2 is paired with 4, 3 is paired with 6, and so on. In symbols, $y = 2 * x$ or $y = 2x$.

function machine In *Everyday Mathematics,* an imaginary device that receives inputs and pairs them with outputs. For example, the function machine on the right pairs an input number with its double. See *function.*

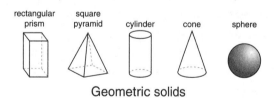

in	out
1	2
2	4
3	6
5	10
20	40
300	600

A function machine and function table

G

geometric solid The surface or surfaces that make up a *3-dimensional* figure such as a prism, pyramid, cylinder, cone, or sphere. Despite its name, a geometric solid is hollow—it does not include the points in its interior. Informally, and in some dictionaries, a solid is defined as both the surface and its interior.

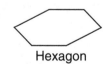

Geometric solids

H

hexagon A 6-sided *polygon.*

Hexagon

Home Link In *Everyday Mathematics,* a suggested follow-up or enrichment activity to be done at home.

horizontal   In a left-to-right orientation. Parallel to the horizon.

I

inch (in.)   A U.S. customary unit of length equal to $\frac{1}{12}$ of a foot and equal to 2.54 centimeters.

K

kite   A *quadrilateral* with two distinct pairs of adjacent sides of equal length. In *Everyday Mathematics,* the four sides cannot all have equal length; that is, a rhombus is not a kite. The diagonals of a kite are perpendicular.

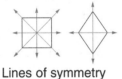
Kite

L

label   A descriptive word or phrase used to put a number or numbers in context. Labels encourage children to associate numbers with real objects. Flags, snowballs, and scary monsters are examples of labels.

line of symmetry   A line that divides a figure into two parts that are reflection images of each other. A figure may have any number of lines of symmetry. For example, the numeral 2 has no lines of symmetry, a square has four lines of symmetry, and a circle has infinitely many lines of symmetry. Also called a symmetry line.

Lines of symmetry

M

measurement unit   The reference unit used when measuring. Examples of basic units include inches for length, grams for mass or weight, cubic inches for volume or capacity, seconds for elapsed time, and degrees Celsius for change of temperature. Compound units include square centimeters for area and miles per hour for speed.

meter (m)   The basic metric unit of length from which other metric units of length are derived. Originally, the meter was defined as $\frac{1}{10,000,000}$ of the distance from the North Pole to the equator along a meridian passing through Paris. From 1960 to 1983, the meter was redefined as 1,630,763.73 wavelengths of orange-red light from the element krypton. Today, the meter is the distance light travels in a vacuum in $\frac{1}{299,792,458}$ second. One meter is equal to 10 decimeters, 100 centimeters, and 1,000 millimeters.

N

Name collection   A collection of equivalent names for a number. A name collection for 5 might include $4 + 1$, $3 + 2$, and so on.

negative numbers   Numbers less than 0; the opposites of the *positive numbers,* commonly written as a positive number preceded by a "−" or *OPP.* Negative numbers are plotted left of 0 on a horizontal line or below 0 on a vertical number line.

number grid   In *Everyday Mathematics,* a table in which consecutive numbers are arranged in rows, usually 10 columns per row. A move from one number to the next within a row is a change of 1; a move from one number to the next within a column is a change of 10.

Number grid

number line   A line on which points are indicated by tick marks that are usually at regularly spaced intervals from a starting point called the origin, the zero point, or simply "0". Numbers are associated with the tick marks on a scale defined by the unit interval from 0 to 1.

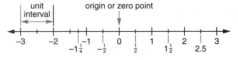
A number line

number model   A *number sentence,* expression, or other representation that models a number story or situation. For example, the story *Sally had $5, and then she earned $8,* can be modeled as the number sentence $5 + 8 = 13$, as the expression $5 + 8$, or by

$$\begin{array}{r} 5 \\ +8 \\ \hline 13, \end{array}$$

number scroll   In *Everyday Mathematics,* a series of *number grids* taped together.

Number scroll

number sentence   Two expressions with a relation symbol. For example,

$$5 + 5 = 10$$
$$2 - ? = 8$$
$$16 \leq a * b$$
$$a^2 + b^2 = c^2$$

Number sentences

numeration   A method of numbering or of reading and writing numbers. In *Everyday Mathematics,* numeration activities include counting, writing numbers, identifying equivalent names for numbers in *name-collection* boxes, exchanging coins such as 5 pennies for 1 nickel, and renaming numbers in computation.

O

octagon   An 8-sided *polygon.*

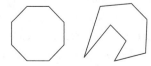

Octagons

operation   A rule performed on one or more mathematical objects such as numbers, variables, or expressions, to produce another mathematical object. Addition, subtraction, multiplication, and division are the four basic arithmetic operations. Taking a square root, squaring a number, and multiplying both sides of an equation by the same number are also operations. In *Everyday Mathematics,* students learn about many operations along with several procedures, or algorithms, for carrying them out.

ordinal number   The position or order of something in a sequence, such as first, third, or tenth. Ordinal numbers are commonly used in dates, as in "May fifth" instead of "May five."

P

parallelogram   A *quadrilateral* with two pairs of parallel sides. Opposite sides of a parallelogram have the same length and opposite angles have the same measure. All rectangles are parallelograms, but not all parallelograms are rectangles because parallelograms do not necessarily have right angles.

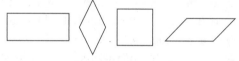

Parallelograms

pattern   A repetitive order or arrangement. In *Everyday Mathematics,* students mainly explore visual and number patterns in which elements are arranged so that what comes next can be predicted.

pattern blocks   A set of *polygon*-shaped blocks of varying sizes in which smaller blocks can be placed on larger blocks to show fractional parts. The blocks are used for geometric shape identification and fraction activities.

Pattern-Block Template   In *Kindergarten Everyday Mathematics,* a sheet of plastic with geometric shapes cut out, used to draw patterns and designs.

pentagon A 5-sided *polygon*.

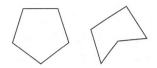

Pentagons

perimeter The distance around the boundary of a *2-dimensional* figure. The perimeter of a *circle* is called its circumference. A formula for the perimeter P of a *rectangle* with length l and width w is $P = 2 * (l + w)$. Perimeter comes from the Greek for "around measure."

pictograph A graph constructed with pictures or symbols.

A pictograph

place value A system that gives a *digit* a value according to its position, or place, in a number. In our standard, base-10 (decimal) system for writing numbers, each place has a value ten times that of the place to its right and one-tenth the value of the place to its left.

A place-value chart

polygon
A *2-dimensional* figure formed by three or more line segments (sides) that meet only at their endpoints (vertices) to make a closed path. The sides may not cross one another.

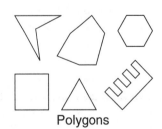

Polygons

polyhedron A *3-dimensional* figure formed by *polygons* with their interiors (faces) and having no holes. Plural is *polyhedrons* or *polyhedra*.

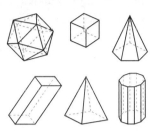

Polyhedrons

positive numbers Numbers greater than 0; the opposites of the *negative numbers*. Positive numbers are plotted right of 0 on a horizontal number line or above 0 on a vertical number line.

Q

quadrangle Same as *quadrilateral*.

quadrilateral A 4-sided *polygon*. See *square, rectangle, parallelogram, rhombus, kite,* and *trapezoid*.

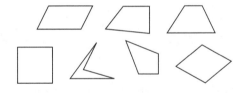

Quadrilaterals

R

rational counting Counting using one-to-one matching. For example, counting a number of chairs, people, crackers, and so on.

rectangle A *parallelogram* with all right angles.

reference frame A system for locating numbers within a given context, usually with reference to an origin or zero point. For example, number lines, clocks, calendars, temperature scales, and maps are reference frames.

rhombus A *parallelogram* with all sides the same length. All rhombuses are parallelograms. Every square is a rhombus, but not all rhombuses are squares. Also called a diamond. Plural is *rhombuses* or *rhombi*.

Rhombuses

S

scale (1) The relative size of something. (2) A tool for measuring *weight*.

second (s) (1) A unit of time defined as $\frac{1}{31,556,925.9747}$ of the tropical year at midnight, Eastern Time on New Year's Day, 1900. There are 60 seconds in a minute. (2) An *ordinal number* in the sequence *first, second, third, ...*

skip counting Rote counting by intervals, such as 2s, 5s, or 10s.

slate A lap-sized (about 8" by 11") chalkboard or whiteboard that children use in *Everyday Mathematics* for recording responses during group exercises and informal group assessments.

speed A rate that compares distance traveled with the time taken to travel that distance. For example, if a car travels 100 miles in 2 hours, then its average speed is $\frac{100 \text{ mi}}{2 \text{ hr}}$, or 50 miles per hour.

square A *rectangle* with all sides of equal length. All angles in a square are right angles.

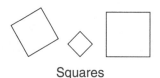

Squares

standard unit A unit of measure that has been defined by a recognized authority, such as a government or a standards organization. For example, inches, meters, miles, seconds, pounds, grams, and acres are all standard units.

survey A study that collects *data*. Surveys are commonly used to study "demographics" such as people's characteristics, behaviors, interests, and opinions.

symmetry The balanced distribution of points over a line or around a point in a symmetric figure. See *line of symmetry*.

A figure with   A figure with
line symmetry   rotation
                    symmetry

T

tally (1) To keep a record of a count, commonly by making a mark for each item as it is counted. (2) The mark used in a count. Also called tally mark and tick mark.

temperature How hot or cold something is relative to another object or as measured on a standardized *scale* such as *degrees Celsius* or *degrees Fahrenheit*.

3-dimensional (3-D) figure A figure whose points are not all in a single plane. Examples include prisms, pyramids, and spheres, all of which have length, width, and height.

timeline   A *number line* showing when events took place. In some timelines the origin is based on the context of the events being graphed, such as the birth date of the child's life graphed below. The origin can also come from another reference system, such as the year A.D., in which case the scale below might cover the years 2000 through 2005.

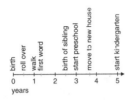

A timeline of a
child's milestones

trapezoid   A *quadrilateral* that has exactly one pair of parallel sides. In *Everyday Mathematics,* both pairs of sides cannot be parallel; that is, a parallelogram is not a trapezoid.

Trapezoids

triangle   A 3-sided polygon.

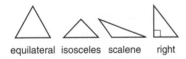

Triangles

2-dimensional (2-D) figure   A figure whose points are all in one plane but not all on one line. Examples include polygons and circles, all of which have length and width but no height.

V

vertical   Upright; perpendicular to the horizon. Compare to *horizontal.*

volume   (1) The amount of space occupied by a *3-dimensional figure.* Same as capacity. (2) Less formally, the amount a container can hold. Volume is often measured in cubic units, such as cm^3, cubic inches, or cubic feet.

W

weight   A measure of how heavy something is; the force of gravity on an object. An object's mass is constant, but it weighs less in weak gravity than in strong gravity. For example, a person who weighs 150 pounds in San Diego weighs about 23 pounds on the moon.

"What's My Rule?" problem   In *Everyday Mathematics,* a problem in which two of the three parts of a *function* (input, output, and rule) are known, and the third is to be found out.

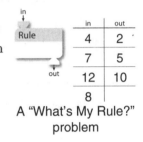

A "What's My Rule?"
problem

Y

yard (yd)   A U.S. customary unit of length equal to 3 feet or 36 inches. To Henry I of England, a yard was the distance from the tip of the nose to the tip of the middle finger. In *Everyday Mathematics,* it is from the center of the chest to the tip of the middle finger.

K–2 Games Correlation Chart

Game	Grade K Activity	Grade 1 Lesson	Grade 2 Lesson	Numeration	Mental Math	Basic Facts	Operations	Patterns	Geometry	Money	Time	Probability	Calculator
Addition Card Draw			12♦5		●	●	●						
Addition Spin		*	4♦2		●	●	●						
Addition Top-It	4♦2	6♦1	1♦4	●	●	●	●						
Animal Weight Top-It		5♦5		●	●	●	●						
Array Bingo			6♦9	●	●	●	●						
Attribute Spinner Game	5♦14								●				
Attribute Train Game		7♦2						●	●				
Base-10 Exchange		5♦3	3♦4	●									
Base-10 Trading Game			6♦5	●		●	●						
Basketball Addition			7♦3		●	●	●						
Beat the Calculator (Addition)		5♦11	2♦2		●	●	●						●
Beat the Calculator (Multiplication)			11♦9		●	●	●						●
Before and After		3♦1	*	●									
Bunny Hop		1♦5		●		●							
Clear the Board	7♦12			●	●	●	●						
Coin-Dice		3♦12		●						●			
Coin Exchange		6♦10	*			●				●			
Coin Top-It		2♦13	1♦4	●						●			
Count and Sit	2♦6			●									
Cover Half	6♦11						●						
Cover the Board	7♦12			●	●	●	●						
Dice Addition and Dice Subtraction	7♦6			●	●	●	●						
Dice Race	3♦3			●								●	
Difference Game		5♦7	2♦12	●	●	●	●			●			
Digit Game		5♦1	3♦1	●	●	●							
Dime-Nickel-Penny Grab		3♦13		●			●			●			
Disappearing Train Game	3♦13			●	●	●	●						
Dollar Rummy			3♦5	●	●	●	●			●			
Domino Concentration	3♦5			●									
Domino Top-It		3♦14	2♦2	●	●	●	●						
Doubles or Nothing			2♦3	●	●	●	●						
Equivalent Fractions Game			8♦5	●									
Fact Extension Game		*	4♦8		●	●	●	●					
Fact Power Game		6♦4			●	●	●						
Find the Block	5♦3								●	●			
Follow the Leader	2♦6			●									
Fraction Top-It			8♦6	●									
Give the Next Number	1♦12			●	●			●					
Go Forward, Back Up	4♦1			●	●								
Growing Train Game	3♦13			●	●	●	●						
Growing and Disappearing Train Game	3♦13			●	●	●	●						
Guess My Number	5♦4			●	●			●					
High Low	7♦14			●								●	
High Roller	8♦4	2♦12	3♦7	●	●	●	●						
Hit the Target		*	7♦2		●		●						●
I Spy	2♦1	10♦5							●	●			
Make My Design		7♦1								●			
Match Up	1♦5			●									

* indicates game only in *My Reference Book*.

Game	Grade K Activity	Grade 1 Lesson	Grade 2 Lesson	Numeration	Mental Math	Basic Facts	Operations	Patterns	Geometry	Money	Time	Probability	Calculator
Matching Coin Game	2◆8									•			
Money Cube	7◆1									•			
Money Exchange Game			1◆5	•						•			
Money Grid	7◆1									•			
Monster Squeeze	3◆6	1◆2		•									
Name that Number		*	2◆9	•	•	•							
Nickel-Penny Grab		2◆11		•			•			•			
Number-Grid Difference		*	2◆12	•		•	•	•					
Number-Grid Game	5◆16	9◆2	1◆8				•	•					
Number-Grid Grab	7◆13			•		•	•	•					
Number-Grid Search	5◆16			•				•					
Number Gymnastics	8◆6			•	•		•						
Number-Line Squeeze		1◆1		•									
Number Top-It		1◆11		•									
One-Dollar Exchange		8◆2	*	•						•			
One Dollar Game	8◆8			•						•			
$1, $10, $100 Exchange Game		10◆4	1◆5	•	•	•	•			•			
Ones, Tens, Hundreds Game	8◆1			•									
Paper Money Exchange Game	8◆1			•						•			
Pattern Cover Up	4◆5							•					
Penny-Dice Game		1◆3		•						•			
Penny-Dime-Dollar Exchange			3◆2	•						•			
Penny-Dime Exchange	6◆7			•						•			
Penny Grab		2◆8		•	•	•				•			
Penny-Nickel-Dime Exchange		5◆13		•						•			
Penny-Nickel Exchange	5◆10	2◆10	1◆5	•						•			
Penny Plate		2◆8	1◆6	•	•					•			
Pick-a-Coin			10◆3							•			•
Plus or Minus Game	7◆12				•		•						
Plus or Minus Steps	7◆12				•		•						
Quarter-Dime-Nickel-Penny Grab		6◆9		•						•			
The Raft Game	5◆10			•									
Read My Mind	6◆12								•				
Rock, Paper, Scissors		1◆8											•	
Rolling for 50		2◆1		•									
Shaker Addition Top-It		4◆12		•	•	•	•						
Soccer Spin			7◆8									•	
Spin a Number	2◆4			•									
Spinning for Money			3◆2	•						•			
Stand Up If...	6◆3								•				
Stick Pick-Up	3◆11											•	
Teen Frame	3◆16			•	•								
Teen Tangle	2◆10			•									
Three Addends			6◆1		•	•	•						
3, 2, 1 Game		8◆5	*		•	•	•	•	•				
Time Match	8◆12	4◆4	*								•		
Top-It	4◆2	1◆6	*	•									
Tricky Teens	2◆10			•									
Tric-Trac		6◆8	*	•	•	•	•						
Two-Fisted Penny Addition			1◆6	•	•	•	•						
Walk Around the Clock	8◆3										•		
"What's My Rule?" Fishing	4◆14								•				
Who Am I Thinking Of?	4◆14								•				

K–2 Grade-Level Goals

Everyday Mathematics organizes content through Program Goals and Grade-Level Goals. For more information, see the *Assessment Handbook*.

The Grade-Level Goals are divided according to the content strands below.

Content Strands   Pages

How to Read the Grade-Level Goals Chart

Each section of the chart includes Grade-Level Goals organized by content strand.

Number and Numeration ●—Content strand name

Content	Kindergarten	Grade 1	Grade 2
Rote counting	1. Count on by 1s to 100; count on by 2s, 5s, and 10s and count back by 1s with number grids, number lines, and calculators. [Number and Numeration Goal 1]	1. Count on by 1s, 2s, 5s, and 10s past 100 and back by 1s from any number less than 100 with and without number grids, number lines, and calculators. [Number and Numeration Goal 1]	1. Count on by 1s, 2s, 5s, 10s, 25s, and 100s past 1,000 and back by 1s, 10s, and 100s from any number less than 1,000 with and without number grids, number lines, and calculators. [Number and Numeration Goal 1] ●

—This column identifies the major mathematical concepts within each content strand.

A complete list of Grade-Level Goals for this grade and the two grades above it demonstrates how the goals evolve from grade to grade.

Grade-Level Goals are numbered for easy identification.

Number and Numeration

Content	Kindergarten	Grade 1	Grade 2
Rote counting	1. Count on by 1s to 100; count on by 2s, 5s, and 10s and count back by 1s with number grids, number lines, and calculators. [Number and Numeration Goal 1]	1. Count on by 1s, 2s, 5s, and 10s past 100 and back by 1s from any number less than 100 with and without number grids, number lines, and calculators. [Number and Numeration Goal 1]	1. Count on by 1s, 2s, 5s, 10s, 25s, and 100s past 1,000 and back by 1s, 10s, and 100s from any number less than 1,000 with and without number grids, number lines, and calculators. [Number and Numeration Goal 1]
Rational counting	2. Count 20 or more objects; estimate the number of objects in a collection. [Number and Numeration Goal 2]	2. Count collections of objects accurately and reliably; estimate the number of objects in a collection. [Number and Numeration Goal 2]	
Place value and notation	3. Model numbers with manipulatives; use manipulatives to exchange 1s for 10s and 10s for 100s; recognize that digits can be used and combined to read and write numbers; read numbers up to 30. [Number and Numeration Goal 3]	3. Read, write, and model with manipulatives whole numbers up to 1,000; identify places in such numbers and the values of the digits in those places. [Number and Numeration Goal 3]	2. Read, write, and model with manipulatives whole numbers up to 10,000; identify places in such numbers and the values of the digits in those places; read and write money amounts in dollars-and-cents notation. [Number and Numeration Goal 2]
Meanings and uses of fractions	4. Use manipulatives to model half of a region or a collection; describe the model. [Number and Numeration Goal 4]	4. Use manipulatives and drawings to model halves, thirds, and fourths as equal parts of a region or a collection; describe the model. [Number and Numeration Goal 4]	3. Use manipulatives and drawings to model fractions as equal parts of a region or a collection; describe the models and name the fractions. [Number and Numeration Goal 3]
Number theory		5. Use manipulatives to identify and model odd and even numbers. [Number and Numeration Goal 5]	4. Recognize numbers as odd or even. [Number and Numeration Goal 4]
Equivalent names for whole numbers	5. Use manipulatives, drawings, and numerical expressions involving addition and subtraction of 1-digit numbers to give equivalent names for whole numbers up to 20. [Number and Numeration Goal 5]	6. Use manipulatives, drawings, tally marks, and numerical expressions involving addition and subtraction of 1- or 2-digit numbers to give equivalent names for whole numbers up to 100. [Number and Numeration Goal 6]	5. Use tally marks, arrays, and numerical expressions involving addition and subtraction to give equivalent names for whole numbers. [Number and Numeration Goal 5]
Equivalent names for fractions, decimals, and percents			6. Use manipulatives and drawings to model equivalent names for $\frac{1}{2}$. [Number and Numeration Goal 6]
Comparing and ordering numbers	6. Compare and order whole numbers up to 20. [Number and Numeration Goal 6]	7. Compare and order whole numbers up to 1,000. [Number and Numeration Goal 7]	7. Compare and order whole numbers up to 10,000; use area models to compare fractions. [Number and Numeration Goal 7]

Operations and Computation

Content	Kindergarten	Grade 1	Grade 2
Addition and subtraction facts	1. Use manipulatives, number lines, and mental arithmetic to solve problems involving the addition and subtraction of 1-digit whole numbers; demonstrate appropriate fluency with addition and subtraction facts within 5. [Operations and Computation Goal 1]	1. Demonstrate appropriate fluency with addition and subtraction facts through 10 + 10. [Operations and Computation Goal 1]	1. Demonstrate automaticity with all addition facts through 10 + 10 and fluency with the related subtraction facts. [Operations and Computation Goal 1]
Addition and subtraction procedures		2. Use manipulatives, number grids, tally marks, mental arithmetic, and calculators to solve problems involving the addition and subtraction of 1-digit whole numbers with 2-digit whole numbers; calculate and compare the values of combinations of coins. [Operations and Computation Goal 2]	2. Use manipulatives, number grids, tally marks, mental arithmetic, paper and pencil, and calculators to solve problems involving the addition and subtraction of multidigit whole numbers; describe the strategies used; calculate and compare values of coin and bill combinations. [Operations and Computation Goal 2]
Computational estimation		3. Estimate reasonableness of answers to basic fact problems (e.g., Will 7+8 be more or less than 10?). [Operations and Computation Goal 3]	3. Make reasonable estimates for whole number addition and subtraction problems; explain how the estimates were obtained. [Operations and Computation Goal 3]
Models for the operations	2. Identify join and take-away situations. [Operations and Computation Goal 2]	4. Identify change-to-more, change-to-less, comparison, and parts-and-total situations. [Operations and Computation Goal 4]	4. Identify and describe change, comparison, and parts-and-total situations; use repeated addition, arrays, and skip counting to model multiplication; use equal sharing and equal grouping to model division. [Operations and Computation Goal 4]

Data and Chance

Content	Kindergarten	Grade 1	Grade 2
Data collection and representation	1. Collect and organize data to create class-constructed tally charts, tables, and bar graphs. [Data and Chance Goal 1]	1. Collect and organize data to create tally charts, tables, bar graphs, and line plots. [Data and Chance Goal 1]	1. Collect and organize data or use given data to create tally charts, tables, graphs, and line plots. [Data and Chance Goal 1]
Data analysis	2. Use graphs to answer simple questions. [Data and Chance Goal 2]	2. Use graphs to answer simple questions and draw conclusions; find the maximum and minimum of a data set. [Data and Chance Goal 2]	2. Use graphs to ask and answer simple questions and draw conclusions; find the maximum, minimum, mode, and median of a data set. [Data and Chance Goal 2]
Qualitative probability	3. Describe events using *certain, possible, impossible*, and other basic probability terms. [Data and Chance Goal 3]	3. Describe events using *certain, likely, unlikely, impossible*, and other basic probability terms. [Data and Chance Goal 3]	3. Describe events using *certain, likely, unlikely, impossible* and other basic probability terms; explain the choice of language. [Data and Chance Goal 3]

Measurement and Reference Frames

Content	Kindergarten	Grade 1	Grade 2
Length, weight, and angles	1. Use nonstandard tools and techniques to estimate and compare weight and length; identify standard measuring tools. [Measurement and Reference Frames Goal 1]	1. Use nonstandard tools and techniques to estimate and compare weight and length; measure length with standard measuring tools. [Measurement and Reference Frames Goal 1]	1. Estimate length with and without tools; measure length to the nearest inch and centimeter; use standard and nonstandard tools to measure and estimate weight. [Measurement and Reference Frames Goal 1]
Area, perimeter, volume, and capacity			2. Partition rectangles into unit squares and count unit squares to find areas. [Measurement and Reference Frames Goal 2]
Units and systems of measurement			3. Describe relationships between days in a week and hours in a day. [Measurement and Reference Frames Goal 3]
Money	2. Identify pennies, nickels, dimes, quarters, and dollar bills. [Measurement and Reference Frames Goal 2]	2. Know and compare the value of pennies, nickels, dimes, quarters, and dollar bills; make exchanges between coins. [Measurement and Reference Frames Goal 2]	4. Make exchanges between coins and bills. [Measurement and Reference Frames Goal 4]
Temperature	3. Describe temperature using appropriate vocabulary, such as *hot*, *warm*, and *cold*; identify a thermometer as a tool for measuring temperature. [Measurement and Reference Frames Goal 3]	3. Identify a thermometer as a tool for measuring temperature; read temperatures on Fahrenheit and Celsius thermometers to the nearest 10°. [Measurement and Reference Frames Goal 3]	5. Read temperature on both the Fahrenheit and Celsius scales. [Measurement and Reference Frames Goal 5]
Time	4. Describe and use measures of time periods relative to a day and week; identify tools that measure time. [Measurement and Reference Frames Goal 4]	4. Use a calendar to identify days, weeks, months, and dates; tell and show time to the nearest half and quarter hour on an analog clock. [Measurement and Reference Frames Goal 4]	6. Tell and show time to the nearest five minutes on an analog clock; tell and write time in digital notation. [Measurement and Reference Frames Goal 6]

Geometry

Content	Kindergarten	Grade 1	Grade 2
Lines and angles			1. Draw line segments and identify parallel line segments. [Geometry Goal 1]
Plane and solid figures	1. Identify and describe plane and solid figures including circles, triangles, squares, rectangles, spheres, and cubes. [Geometry Goal 1]	1. Identify and describe plane and solid figures including circles, triangles, squares, rectangles, spheres, cylinders, rectangular prisms, pyramids, cones, and cubes. [Geometry Goal 1]	2. Identify, describe, and model plane and solid figures including circles, triangles, squares, rectangles, hexagons, trapezoids, rhombuses, spheres, cylinders, rectangular prisms, pyramids, cones, and cubes. [Geometry Goal 2]
Transformations and symmetry	2. Identify shapes having line symmetry. [Geometry Goal 2]	2. Identify shapes having line symmetry; complete line-symmetric shapes or designs. [Geometry Goal 2]	3. Create and complete two-dimensional symmetric shapes or designs. [Geometry Goal 3]

Patterns, Functions, and Algebra

Content	Kindergarten	Grade 1	Grade 2
Patterns and functions	1. Extend, describe, and create visual, rhythmic, and movement patterns; use rules, which will lead to functions, to sort, make patterns, and play "What's My Rule?" and other games. [Patterns, Functions, and Algebra Goal 1]	1. Extend, describe, and create numeric, visual, and concrete patterns; solve problems involving function machines, "What's My Rule?" tables, and Frames-and-Arrows diagrams. [Patterns, Functions, and Algebra Goal 1]	1. Extend, describe, and create numeric, visual, and concrete patterns; describe rules for patterns and use them to solve problems; use words and symbols to describe and write rules for functions involving addition and subtraction and use those rules to solve problems. [Patterns, Functions, and Algebra Goal 1]
Algebraic notation and solving number sentences	2. Read and write expressions and number sentences using the symbols $+$, $-$, and $=$. [Patterns, Functions, and Algebra Goal 2]	2. Read, write, and explain expressions and number sentences using the symbols $+$, $-$, and $=$, and the symbols $>$ and $<$ with cues; solve equations involving addition and subtraction. [Patterns, Functions, and Algebra Goal 2]	2. Read, write, and explain expressions and number sentences using the symbols $+$, $-$, $=$, $>$, and $<$; solve number sentences involving addition and subtraction; write expressions and number sentences to model number stories. [Patterns, Functions, and Algebra Goal 2]
Properties of arithmetic operations		3. Apply the Commutative and Associative Properties of Addition and the Additive Identity to basic addition fact problems. [Patterns, Functions, and Algebra Goal 3]	3. Describe the Commutative and Associative Properties of Addition and the Additive Identity and apply them to mental arithmetic problems. [Patterns, Functions, and Algebra Goal 3]

Scope and Sequence Chart

Throughout *Everyday Mathematics*, children repeatedly encounter skills in each of the content strands. Each exposure builds on and extends children's understanding. They study important concepts over consecutive years through a variety of formats. The Scope and Sequence Chart shows the units in which these exposures occur. The symbol ● indicates that the skill is introduced or taught. The symbol ■ indicates that the skill is revisited, practiced, or extended. These levels refer to unit content within the *K–6 Everyday Mathematics* curriculum.

The skills are divided according to the content strands below.

Content Strands

How to Read the Scope and Sequence Chart

Each section of the chart includes a content strand title, three grade-level columns divided by units or sections, and a list of specific skills grouped by major concepts.

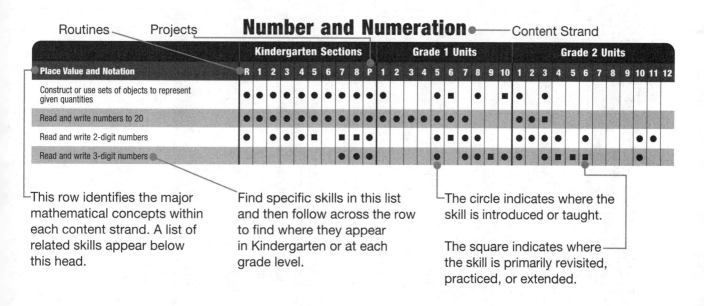

This row identifies the major mathematical concepts within each content strand. A list of related skills appear below this head.

Find specific skills in this list and then follow across the row to find where they appear in Kindergarten or at each grade level.

The circle indicates where the skill is introduced or taught.

The square indicates where the skill is primarily revisited, practiced, or extended.

Number and Numeration

	Kindergarten Sections									Grade 1 Units											Grade 2 Units											
Rote Counting	R	1	2	3	4	5	6	7	8	P	1	2	3	4	5	6	7	8	9	10	1	2	3	4	5	6	7	8	9	10	11	12
Perform rote counting	●	●	●	●	●	■	■	●	■	●	●	●	●	●	●	●	●	●	●	■	●	●	●	●	●	●	●			●	●	■
Count by 2s, 5s, and 10s forward and backward (may include the use of concrete objects)	●		●	●	■	●	■	■	●	●	●	●	●	●	●	■	■	●	●	■	■	●	●	■	■	■	■	■		●	●	■
Count backward from 10 to 1			●		■	■		■	■					●	■																	
Count by numbers greater than 10						●							●			●		●		●	■	■	●							●	●	
Count by 25s													●							●			●									
Count by 100s					■	●	■	■	●			●	●		●		■	●	■	●	●	■		■			■	●		●		■
Count up and back on a number grid						●										●					●	●										
Relate counting to addition and subtraction					■		■		■				●				■	■		■	●	●		■		■						
Locate numbers on a number line; count up and back on a number line; complete a number line	●			●	●			■		●			■	●					■		●	●		■								
Count using a calculator or calculator repeat key				●	●		■		●		●	●	●		●	■		●		■	●	■				■	●	■	■	●	■	■
Rational Counting	R	1	2	3	4	5	6	7	8	P	1	2	3	4	5	6	7	8	9	10	1	2	3	4	5	6	7	8	9	10	11	12
Perform rational counting	●	●	●	●	●	●	●	●	●	●	●	●	●	●	●	●	●	■	●	●	●	●	●	●	●	●	●	■	●	●	●	
Compare number of objects in sets of concrete objects	●	●	●	■	●	●	■	■	●	●	●		■													●	●	■	●	●		
Estimate quantities of objects				■	●	■	●	■		●	●			■	●			●			●	●										
Place Value and Notation	R	1	2	3	4	5	6	7	8	P	1	2	3	4	5	6	7	8	9	10	1	2	3	4	5	6	7	8	9	10	11	12
Construct or use sets of objects to represent given quantities	●	●	●	●	●	■	■	●	●	●	●	●			●	■		●		■	●	●		●	●							
Read and write numbers to 20	●	●	●	●	●	●	●	●	●	●	●	●					●			●	●	●				●						
Read and write 2-digit numbers	●	●	●		●	■	■	●	■	●				■			●		●	●	●	●				●	●					
Read and write 3-digit numbers										●								●	■	●	●	●		●	■	■			●	●	●	
Read and write 4- and 5-digit numbers																					●	●		●					●	●	●	
Display and read numbers on a calculator	●				●	●	●	●	●	●	●			●	●					■	●	●			■				●	●		●

Place Value and Notation (cont.)

	R	1	2	3	4	5	6	7	8	P	1	2	3	4	5	6	7	8	9	10	1	2	3	4	5	6	7	8	9	10	11	12
Use multimedia and technology to explore number concepts	●				●	●	●		●									●														
Read, write, or use ordinal numbers		●				●	●															■				■			●			
Name the ordinal positions in a sequence and "next" and "last" positions		●																														
Identify the number that is one more or one less than a given number	●	■ ■	■										■						● ■					●								
Explore place value using a number grid	●					●		●		●		●			●			●														
Identify place value in 2-digit numbers								●	●		■	● ■	●	●	■	● ■		● ■	●	●	■		●	●	■	●			●	●		■
Identify place value in 3-digit numbers									■		●		●	●		■		●		●	●	●	●	●	■			■		●		■
Identify place value in 4-digit numbers																			●							●	●	● ■	● ■	● ■		■
Identify place value in larger numbers																							●									
Make exchanges among place values	●					●		●	●		●		●	●				●		●				●					●			
Make least and greatest numbers with randomly selected digits					●				■											■	●				■							
Write numbers in expanded notation																																
Use cents notation											●	●	●			■	●				●									●		
Use dollars-and-cents notation									●		● ●	● ●	●		■	● ■	●	●		■		●	■	●	●			● ■	● ■	●		
Use calculator to count/compute money amounts																		■			■											
Explore uses for decimals																	●													●		

Meanings and Uses of Fractions

	R	1	2	3	4	5	6	7	8	P	1	2	3	4	5	6	7	8	9	10	1	2	3	4	5	6	7	8	9	10	11	12
Understand the meaning or uses of fractions							●	■	■								■	●	●	■								●	●	■		
Construct concrete models of fractions and equivalent fractions; identify fractions on a number line																	●	●							●			●	■	■		
Identify pennies and dimes as fractional parts of a dollar																			●												●	■
Identify numerator and denominator																		●									●	●		●		

Scope and Sequence Chart   **105**

	Kindergarten Sections									Grade 1 Units											Grade 2 Units											
	R	1	2	3	4	5	6	7	8	P	1	2	3	4	5	6	7	8	9	10	1	2	3	4	5	6	7	8	9	10	11	12
Meanings and Uses of Fractions (cont.)																																
Shade and identify fractional parts of a region							●										■	●	●	■								●		●		■
Shade and identify fractional parts of a set								■								■	●	●	■									●	●		■	■
Understand that the amount represented by a fraction depends on the size of the whole (ONE)																		■														■
Use fractions in number stories									■										■													
Number Theory																																
Explore or identify even and odd numbers						■						●							■								●		●			
Equivalent Names for Whole Numbers																																
Find equivalent names for numbers	●									●	●	●	●	●	●	●	●	●	●	●	●	●	●	●	●	●	●	●	●	●	●	●
Use Roman numerals											■																					
Equivalent Names for Fractions, Decimals, and Percents																																
Find equivalent fractions																			●										■			
Comparing and Ordering Numbers																																
Compare and order numbers to 20	●			●	●	●	■	■		●	●	●	●	●	●	●	■	■	■	■	●		■				■		●			●
Compare and order 2-digit numbers	●	●	●	●	●	●	■	■			●	●	●	●	●	●	■	■	■	●	●		●				●					●
Compare and order 3-digit numbers															●				■				●							●		
Compare and order 4- or 5-digit numbers																								■								
Compare and order larger numbers																														●	■	
Compare numbers using the symbols <, >, and =	●							■		●	●	●	●	●	●	●	■	■	●	■	●	●				●		●	●	■	■	■
Compare and order fractions; use manipulatives to identify/compare fractions																														■		
Compare fractions less than one																												●				

Operations and Computation

	Kindergarten Sections										Grade 1 Units										Grade 2 Units											
Addition and Subtraction Facts	R	1	2	3	4	5	6	7	8	P	1	2	3	4	5	6	7	8	9	10	1	2	3	4	5	6	7	8	9	10	11	12
Find/use complements of 10	■							●			●		■	■	■	■			■							■	●		■	●	■	
Practice basic facts; know +/− fact families			●	●	●	●	●	●	●	●	●	●	●	●	●	●	●	■	■	■	●	●	■	■	●	●	■	■	■	●	■	●
Practice extensions of basic facts					■														■					■								
Make and solve number-grid puzzles													●						●	●	■	●										
Addition and Subtraction Procedures	R	1	2	3	4	5	6	7	8	P	1	2	3	4	5	6	7	8	9	10	1	2	3	4	5	6	7	8	9	10	11	12
Understand meaning of addition/subtraction; model addition/subtraction using concrete objects	■	●	●	●	●	■	■	●	●	●	●	●	●	●	●	●	●	■	●	●	●	●	●	●	●		●					●
Investigate the inverse relationships between addition and subtraction		●	●	●	●	■	■	●	●		●	●	●	●	●	●	●	●	●	■	●	●	■	●	■			■		■		
Use mental arithmetic or fact strategies to add/subtract				●	●	●	●	●	●	●	●	●	●	●	●	●	●	●	●	●	●	●	■	●	■	●	●	■	■	■	■	■
Use addition to find the total number of objects in rectangular arrays																			●			●		■	●				●		●	
Use addition/subtraction algorithms																				■	●	●	■	●	●	●	●	●	●	●	●	■
Explore calculator functions	●					●	●		■						●	●		■	●	●	●	●	■	■	●	●	■	■	■			●
Make up and/or solve 1- or 2-step addition/subtraction number stories; determine operation needed to solve a problem		●	●	●	●	●	●	●	●	●	●	●	●	●	●	●	●	●	●	●	●	●	●	●	●	●	●	●	●	●	●	●
Use an Addition/Subtraction Facts Table																			●			●										
Determine the value of the unknown number in an addition or subtraction problem					●						●	●	●					■	●	■	●	●	●	●	●	●	●		●		●	
Add/subtract using a number grid															■	■			●	■	●	●	●	■	●	●	●			■	■	
Add/subtract using a number line	●				●								●								●	●		●			●	●	■		■	●
Add/subtract using a calculator							■		●										■		●	●		●	●		●			■		●
Add/subtract multiples of 10									●											●		■		■		■			■	●		

Operations and Computation (cont.)

	Kindergarten Sections										Grade 1 Units										Grade 2 Units											
	R	1	2	3	4	5	6	7	8	P	1	2	3	4	5	6	7	8	9	10	1	2	3	4	5	6	7	8	9	10	11	12
Addition and Subtraction Procedures (cont.)																																
Add 3 or more 1-digit numbers												●	●											●	●	●	■	■	■	■	●	●
Add/subtract 2-digit numbers													●		●	●	●	●	●	●	●			●	●	●	●	■	■	■	●	●
Add 3 or more 2-digit numbers															●									●	●	●					●	●
Add/subtract 3- and 4-digit numbers																	●		●	●				●			■		●		●	●
Add/subtract money amounts/decimals; make change																	●		●	●	●			■	●	■	●		■		●	■
Solve money number stories															●	●		●	●	●				●		■		■			●	
Make change																						■		■	■				■			■
Multiplication and Division Facts																											●				■	●
Practice multiplication/division facts																				●							●				■	●
Find complements for multiples of 10																											●					
Multiplication and Division Procedures																																
Use manipulatives, drawings/arrays, number sentences, repeated addition, or story problems to explain and demonstrate the meaning of multiplication/division							●							●							●				■		■	●	■	■		●
Understand meaning of multiplication/division and related vocabulary																								●	●	●	●				●	
Make up and/or solve multiplication/division number stories																									●	●	●	●			●	●
Investigate relationships between multiplication and division																										●					●	
Multiply/divide using a number line or number grid																										■					●	
Explore square numbers																															■	
Use a calculator to multiply or divide																															●	
Use a Multiplication/Division Facts Table																															●	●

Multiplication and Division Procedures (cont.)

	R	1	2	3	4	5	6	7	8	P	1	2	3	4	5	6	7	8	9	10	1	2	3	4	5	6	7	8	9	10	11	12
Use mental arithmetic to multiply/divide																														●	●	■
Identify factors of a number																															●	

Computational Estimation

	R	1	2	3	4	5	6	7	8	P	1	2	3	4	5	6	7	8	9	10	1	2	3	4	5	6	7	8	9	10	11	12
Estimate reasonableness of answers to basic facts													●			■	■		■	●				■	■				●		●	
Use estimation strategies to add/subtract; make ballpark estimates																				●				●	■	■				●	●	
Round whole numbers to the nearest ten																●		●	●													
Estimate costs																■		■						●							●	

Models for Operations

	R	1	2	3	4	5	6	7	8	P	1	2	3	4	5	6	7	8	9	10	1	2	3	4	5	6	7	8	9	10	11	12
Solve change-to-more and change-to-less number stories/diagrams		●	●	●	●	●	■	●	●			●		●			●		●	●		●	●	●	●	●				●		
Solve parts-and-total number stories/diagrams		●	●	●	●	●	■	●	●					●			●		●			●	●	●	●	●		■	■	■	●	■
Solve comparison number stories/diagrams					●										●	●	●					●				●				●	●	●
Solve equal-grouping and equal-sharing division problems					●													■								■				■	■	■

Data and Chance

	Kindergarten Sections										Grade 1 Units										Grade 2 Units											
Data Collection and Representation	R	1	2	3	4	5	6	7	8	P	1	2	3	4	5	6	7	8	9	10	1	2	3	4	5	6	7	8	9	10	11	12
Collect data by counting	●	●		●		●	●	●	■	●	■							■	●			●				●	●			●		●
Collect data by interviewing	●						●												■	■					■	●			■	●		
Collect data by measuring														●					●	●												
Collect data from print sources and/or posters																													■	■		●
Collect data from a map																									■							
Use a weather map									■											●									●			
Conduct a survey	●						●		■																	●						

Data and Chance (cont.)

	Kindergarten Sections									Grade 1 Units											Grade 2 Units											
Data Collection and Representation (cont.)	R	1	2	3	4	5	6	7	8	P	1	2	3	4	5	6	7	8	9	10	1	2	3	4	5	6	7	8	9	10	11	12
Make a tally chart or frequency table	●																				■											■
Record data in a table/chart	●		■		●	■	●	●	■	●	●	●	●	●	●	●	■	■	■	●		●			■	●	●	■			■	●
Record days/events on a timeline	●				●	●				●															■							●
Create/interpret a bar graph, pictograph (picture graph), or Venn diagram	●		■		●	■	●	●	■	●	●	●	●	●	●	●	■	■	■	●		●			■	●	●	■			■	●
Create/interpret a line plot	●						●	●	■	●					■	●				●	■						●		●	●	●	●
Explore graphing software to make a bar graph or line plot																				●						●						
Data Analysis	R	1	2	3	4	5	6	7	8	P	1	2	3	4	5	6	7	8	9	10	1	2	3	4	5	6	7	8	9	10	11	12
Read tables, graphs, and maps (including map scale, scale drawing)	●		■	●	●	●	●	●	■	●	●	●	●	●	●	■	■	■	■	●	■	●			■	●	■	■	●	●	■	●
Summarize and interpret data	●		■	●	●	●	●	●	■	●	●	●	●	●	●	●	■	■	■	●	●	●			■	●	●	●	●	●	●	●
Compare two sets of data; use calculator to compare data	●		■		●			●	■	●	●									●							■					●
Make predictions about data	●		■						■	●	●																					
Identify "more" or "less" from pictographs and bar graphs	●		■						■	●					●					●			●					■				●
Compare quantities from a bar graph	●		●					●	●	●					●	●	●			●	●	●	●			●	●	■				●
Find the minimum/maximum of a data set														●					■								●					●
Find the range														●	●					●	●	●					●					●
Find the median																				●						■	●					●
Find the mode																				●						■	●					●
Use data in problem solving	●																				■											●
Qualitative and Quantitative Probability	R	1	2	3	4	5	6	7	8	P	1	2	3	4	5	6	7	8	9	10	1	2	3	4	5	6	7	8	9	10	11	12
Understand the language of probability to discuss																																

Qualitative and Quantitative Probability (cont.)

Length/Probability	R	1	2	3	4	5	6	7	8	P	1	2	3	4	5	6	7	8	9	10	1	2	3	4	5	6	7	8	9	10	11	12
Explore equal-chance events					■				■						●	●		■														
Participate in games or activities based on chance		●	●	●	●	●	■	■	●	●																						
Predict outcomes; solve problems involving chance outcomes			●	●							●	●	■	■	■	■	■		●	●		●	●									
Conduct experiments; test predictions using concrete objects				●							●																■					
Find combinations (Cartesian products)	●	●	●		■			■	■																			■				

Measurement and Reference Frames

Length, Weight, and Angles	Kindergarten Sections										Grade 1 Units										Grade 2 Units											
	R	1	2	3	4	5	6	7	8	P	1	2	3	4	5	6	7	8	9	10	1	2	3	4	5	6	7	8	9	10	11	12
Name tools used to measure length		●	●	●	●	●	■	■	■	●	●	●	●	●		●	●		●	●				●			■		●			
Estimate, compare, and order lengths/heights of objects	●	■	●	●	■	●	■	■	■	●	●	●	■	●	●	●	●		●	●						■	●	■	●	■		
Compare lengths indirectly	●	●				■																										
Measure lengths with nonstandard units		●	●	●	■	●	■	■	■	●			●	●										■				■				
Measure to the nearest foot						●								●										●				■	●			
Measure to the nearest inch														●	■				●								●		●			
Measure to the nearest ½ inch														●													●					
Investigate the yard																								●		■			●			
Measure to the nearest yard																								●					●			
Measure to the nearest centimeter																●								●	■	■	●	■	●	■	■	■

Measurement and Reference Frames (cont.)

	Kindergarten Sections										Grade 1 Units										Grade 2 Units											
Length, Weight, and Angles (cont.)	R	1	2	3	4	5	6	7	8	P	1	2	3	4	5	6	7	8	9	10	1	2	3	4	5	6	7	8	9	10	11	12
Investigate the meter																●													●			
Measure to the nearest meter and/or decimeter																													●		■	
Solve length/height number stories									●																	●	●	●	■		■	
Investigate the mile and/or kilometer																													●		■	
Use words to describe distance										●																						
Estimate and compare distances									●																			■	●		■	
Solve distance number stories									●	●									■					■			■		●			
Estimate, compare, and order weights			●						●													■							●			
Name tools used to measure weight			●						●	●											●								●			
Order objects by weight											●																					
Use a pan balance			●		■				●	●	●			●	●	●			■					■			●		●			
Use a bath scale										●																						
Use a spring scale												●																				
Choose the appropriate scale																													●			
Solve weight number stories																								■				■	●			

	Kindergarten Sections										Grade 1 Units										Grade 2 Units											
Area, Perimeter, Volume, and Capacity	R	1	2	3	4	5	6	7	8	P	1	2	3	4	5	6	7	8	9	10	1	2	3	4	5	6	7	8	9	10	11	12
Investigate area															●	●								●					●			
Find the area of regular shapes concretely															●	●													●			■
Find the perimeter of regular shapes concretely, graphically, or with pictorial models																												■	●	●	■	■
Find the area of a rectangular region divided into square units																								■					●	●		
Partition rectangles into same-size squares; count to find the total																								■					●			

Area, Perimeter, Volume, and Capacity (cont.)

Skill	R	1	2	3	4	5	6	7	8	P	1	2	3	4	5	6	7	8	9	10	1	2	3	4	5	6	7	8	9	10	11	12
Find the area of irregular shapes concretely																												■	●			
Find the perimeter of irregular shapes concretely, graphically, or with pictorial models																												■	●		■	
Estimate area																													●			
Estimate perimeter																													●			
Compare perimeter and area																												■	●			
Name tools used to measure area																													●			
Estimate volume/capacity			●																●									●				
Name tools used to measure volume and/or capacity										●																			■			
Find volume																												●				
Measure capacities of irregular containers																													■			
Compare and order the capacities of containers			●																●										●			

Units and Systems of Measurement

Skill	R	1	2	3	4	5	6	7	8	P	1	2	3	4	5	6	7	8	9	10	1	2	3	4	5	6	7	8	9	10	11	12
Select and use appropriate nonstandard units to measure time							●																						■			
Estimate the duration of a minute								●	●		●	●																				
Investigate the duration of an hour									●			●																				
Investigate 1-minute intervals																															■	■
Identify equivalent customary units of length																			●										●			
Identify equivalent metric units of length																													●			
Identify customary and/or metric units of weight																													●			
Identify equivalent customary units of weight																													●			
Identify customary and/or metric units of capacity																			●										●			
Identify equivalent customary/metric units of capacity																													●			
Choose the appropriate unit of measure																													●			

Measurement and Reference Frames (cont.)

	Kindergarten Sections										Grade 1 Units										Grade 2 Units											
Money	R	1	2	3	4	5	6	7	8	P	1	2	3	4	5	6	7	8	9	10	1	2	3	4	5	6	7	8	9	10	11	12
Recognize pennies and nickels	●	●		■			●	●	■		●		■		■			■	■		●	■		●	■	■	■	■	■	■	■	■
Recognize dimes	●	●		■		■	●	●	■		●		■	●			■	●	●		●	■		●	■	■	■	■	■	■	■	■
Recognize quarters	●			■		■			■					●	●	●	■	●	●		●	■		●				●	●	●		■
Recognize dollars						■		●	●				●	■		●		●			●	■	●	■					●	●		■
Calculate the value of coin combinations							●	●													●	■		●			●					
Calculate the value of bill combinations									●						■		●	●		■	●	■		■								●
Calculate the value of coins/bills													●											●								
Compare values of sets of coins or money amounts using <, >, and = symbols								■			■	■		●	●	●	●	●	●	■			●			●	■		●	●		
Identify equivalencies and make coin exchanges							●		●		●		●	●	●	●	●	●		■	●			●	●							
Identify equivalencies and make coin/bill exchanges									●											■				●			■			■		
Temperature	R	1	2	3	4	5	6	7	8	P	1	2	3	4	5	6	7	8	9	10	1	2	3	4	5	6	7	8	9	10	11	12
Compare situations or objects according to temperature	●								■					●	●		●		●	■	●	■		●		■	■					
Use a thermometer	●								■		●	●	■	●	■	●	●	●	■	●	●	■		●	■	■	■					
Use the Fahrenheit temperature scale	●										●			●						●	●		●						■			●
Use the Celsius temperature scale	●																			●	●							●		■		●
Solve temperature number stories														●					■					●				■				
Time	R	1	2	3	4	5	6	7	8	P	1	2	3	4	5	6	7	8	9	10	1	2	3	4	5	6	7	8	9	10	11	12
Demonstrate an understanding of the concepts of time; estimates and measures the passage of time using words like before, after, yesterday, today, tomorrow, morning, afternoon, hour, half-hour	●					●	●	●	●		●	●	●	●	●		●				●			■						●		
Order or compare events according to duration; calculate elapsed time	●					●	●	●				●								●	■		●	■			■	■				●
Name tools used to measure time	●						●	●	●												●		■	■								■

Time (cont.)

	R	1	2	3	4	5	6	7	8	P	1	2	3	4	5	6	7	8	9	10	1	2	3	4	5	6	7	8	9	10	11	12
Relates past events to future events							●																									
Investigate A.M. and P.M.	●											●	■								●	■	■	■		■					●	
Name the seasons of the year						●				●																						
Use the calendar; identify today's date	●										●	■	■								●	■	●	■		■						●
Number and name the months in a year or days in the week	●										●		■								●			■		■			■			●
Investigate the second hand; compare the hour and minute hands									●			●	●								●		●									
Use an analog or digital clock to tell time on the hour	●								●		●							●		●	●	■	●	■	●	■			■			●
Tell time on the half-hour											●	■	■	●	●	●	■	■	●	●	■	■	●	●	●	■	■		■			●
Tell time on the quarter-hour																				●	●	■	●	●	●	●	●	■	■	■	■	
Tell time to the nearest 5 minutes														●	●	●	●	●	●	●	●	■	●			■						●
Use digital notation*								●								●	●	●	●	●	●	■	●	■	●			■	■	■	■	■
Tell time to the nearest minute*												■								●	●	■	●	■	■			■	■	■	■	■
Read time in different ways and/or identify time equivalencies								●								■										■						●
Solve time number stories																																

Coordinate Systems

	R	1	2	3	4	5	6	7	8	P	1	2	3	4	5	6	7	8	9	10	1	2	3	4	5	6	7	8	9	10	11	12
Find and name locations with simple relationships on a coordinate system																												■				

* In Grade 2, children record the start time at the top of journal pages on a daily basis.

Geometry

	Kindergarten Sections										Grade 1 Units											Grade 2 Units											
Lines and Angles	R	1	2	3	4	5	6	7	8	P	R	1	2	3	4	5	6	7	8	9	10	1	2	3	4	5	6	7	8	9	10	11	12
Identify and name line segments																										●						■	
Draw line segments with a straightedge																	●									●		■	■				
Draw line segments to a specified length																	■								■				■	■			
Draw designs with line segments																																	
Identify and name points																									●	●							
Model parallel lines on a geoboard																										●							
Draw parallel lines with a straightedge																										●	■						
Identify parallel, nonparallel, and intersecting line segments																										●				■			

	Kindergarten Sections										Grade 1 Units											Grade 2 Units											
Plane and Solid Figures	R	1	2	3	4	5	6	7	8	P	R	1	2	3	4	5	6	7	8	9	10	1	2	3	4	5	6	7	8	9	10	11	12
Explore shape relationships	●			●	●		●	■				●	■				●	■		■	●	■	■			●				●	■		■
Recognizes open and closed figures	●		●		●	●	●	●				●														●							
Identify characteristics of 2-dimensional shapes; sort shapes by attributes	●		■	●	●	●	●	■					●				●	●	■	●	●	●		■	■	●							
Distinguish between defining and non-defining attributes							■										■																
Explore 2-D shapes utilizing technology or multimedia resources																																	
Identify characteristics and use appropriate vocabulary to describe properties of 2-dimensional shapes	●		●			●	●	●	●			●	●			●	●	●	■	●	●	■	■		■	●				●	●		
Construct models of polygons using manipulatives such as straws or geoboards																					●					●					●		
Match objects to outlines of shapes (on a Pattern-Block Template)										■		■														●							
Draw 2-dimensional shapes (such as triangles and quadrilaterals); draw/describe objects in the environment that depict geometric figures	■		●	●								●	●					●	●			■		●		●				■	●	●	■

Plane and Solid Figures (cont.)

Skill	R	1	2	3	4	5	6	7	8	P	1	2	3	4	5	6	7	8	9	10	1	2	3	4	5	6	7	8	9	10	11	12
Combine shapes and take them apart to form other shapes	●		●		●			●	●				●				●		■	●					●			●		●		
Record shapes or designs	■	●	●		●									●							●											
Identify and draw congruent or similar shapes																		■	■									■				
Classify and name polygons			●	■	■		●		●		●	■								■					●						■	■
Compare 2-dimensional shapes					●	●		●	●		●						●								●							■
Compare polygons and non-polygons								■												●					●							
Solve 2-dimensional shapes problems													■									■		■	■							
Decompose shapes into shares						●		●	●	●								■	■	■								●		■	■	
Identify/compare 3-dimensional shapes; sort shapes and/or describe attributes of each group						●	●	●	■	■	●							●	●	●				■	●	●	■			■	■	■
Construct 3-dimensional shapes							●	●	■	■	●								■	●					●				●			
Locate 2-D shapes on 3-D objects; compare 2- and 3-D shapes				●				●																	●							
Explore 3-D shapes utilizing technology							■																									
Identify the number of faces, edges, vertices, and bases of prisms and pyramids																			■								■	■				
Identify the shapes of faces							●	■																								
Explore slanted 3-dimensional shapes																																

Transformations and Symmetry

Skill	R	1	2	3	4	5	6	7	8	P	1	2	3	4	5	6	7	8	9	10	1	2	3	4	5	6	7	8	9	10	11	12
Identify symmetrical figures or symmetry in the environment	●				■		■		■	●					●		●		●	●				●		●	●			●		
Fold and cut symmetrical shapes	●				■		■		■	●					●		●	■		●						●	●		●			
Create/complete a symmetrical design/shape using concrete models, geoboard, and/or technology		●													●		●		●	●						●						
Identify lines of symmetry															●		●		●	●						●	●	■	■	●		
Use objects to explore slides, flips, and turns; predict the results of changing a shape's position or orientation using slides, flips, and turns					●								●											■								

Geometry (cont.)

Spatial	Kindergarten Sections										Grade 1 Units										Grade 2 Units											
	R	1	2	3	4	5	6	7	8	P	1	2	3	4	5	6	7	8	9	10	1	2	3	4	5	6	7	8	9	10	11	12
Recognize that the quantity remains the same when the spatial arrangement changes			●								●																						
Arrange or describe objects by proximity, position, or direction using words such as *over, under, above, below, inside, outside, beside, in front of, behind*			●	■				●	■	●			●				●												■			
Give or follow directions for finding a place or object			●							●							■	●														
Identify left hand and right hand			●							●														■								
Identify structures from different views or match views of the same structures portrayed from different perspectives							●									●								■								
Use objects to explore slides, flips, and turns; predict the results of changing a shape's position or orientation using slides, flips, and turns					●																				■							

Patterns, Functions, and Algebra

Patterns and Functions	Kindergarten Sections										Grade 1 Units										Grade 2 Units											
	R	1	2	3	4	5	6	7	8	P	1	2	3	4	5	6	7	8	9	10	1	2	3	4	5	6	7	8	9	10	11	12
Identify, extend, and create patterns of sounds, physical movement, and concrete objects		●	●	●	●	●	●	●	■	●	●		●	●	●	●	●						●	●	●	■	●	●	●			
Verbally describe changes in various contexts		●		■			●	●				●			●	●	●															
Explore and extend visual patterns		●	●	●	●	●	●	●	■		●		●	●	●	●	●									■	●					
Find patterns and common attributes in objects/people in the real world		●	●	●	●	●	●	●		●	■	■	●	●	●	■	■							●	●		●	●				
Create and complete patterns with 2-dimensional shapes	●				●			●			■		●	●	■	■	■				■		■		■		■		●	■		
Identify and use patterns on a number grid			●					●				■	■		●	●	●	■		●	■	■			■				●	●		

Patterns and Functions (cont.)	R	1	2	3	4	5	6	7	8	P	1	2	3	4	5	6	7	8	9	10	1	2	3	4	5	6	7	8	9	10	11	12
Add and subtract using a number grid																			●		●	■			■					■	■	
Investigate even and odd number patterns; create, describe, extend simple number patterns/sequences			●			●	●	●	●				●	■							●	●	■									
Explore counting patterns using a calculator							●						●								●					■	●					
Solve "What's My Rule?" (e.g. function machine) problems					●			●	●						●	●		■	■	■		●	■	●			●			■	●	■
Solve Frames-and-Arrows problems with one or two rules													●	■	■	■			■	■	■	■	●	■		■		■	■	■	■	
Find patterns in addition and subtraction facts															●	●		●	■			■	●							■	■	
Explore patterns in doubling or halving numbers															●							●				■	●			■	■	
Find patterns in multiplication and division facts																											■				●	●
Find patterns in multiples of 10, 100, and 1,000																							■							■	●	
Algebraic Notation and Solving Number Sentences	**R**	**1**	**2**	**3**	**4**	**5**	**6**	**7**	**8**	**P**	**1**	**2**	**3**	**4**	**5**	**6**	**7**	**8**	**9**	**10**	**1**	**2**	**3**	**4**	**5**	**6**	**7**	**8**	**9**	**10**	**11**	**12**
Determine whether equations are true or false													●	■	●			■														
Use symbols ×, ÷, =																										●	●				●	●
Use symbols +, −, =; pictures; manipulatives; and models to organize, record, and communicate mathematical ideas					●	■	■	●	●				●	●	■	●	●	●	●	●	●	●	■	●	●	■	■	■		●	●	●
Compare numbers using <, > symbols															●	■		■		■	●	■	●	■		■	■		■	■	■	■
Write/solve addition and subtraction number sentences					●	■	■	●	●			●					●	●	●	■	●	■	●	●	■	■	■		●	●	●	
Write/solve number sentences with missing addends									●						●							●				■	■		●	●	●	
Write and solve multiplication number sentences																										●	●				●	●
Write and solve division number sentences																															●	●
Write and solve number sentences with missing factors; know that symbols can be used to represent missing or unknown quantities									●																						●	●

Patterns, Functions, and Algebra (cont.)

	Kindergarten Sections										Grade 1 Units										Grade 2 Units											
	R	1	2	3	4	5	6	7	8	P	1	2	3	4	5	6	7	8	9	10	1	2	3	4	5	6	7	8	9	10	11	12
Order of Operations																																
Make up and/or solve number sentences involving parentheses																														●		
Properties of Arithmetic Operations																																
Investigate properties of addition/subtraction			●	●	●	■	■	●	●		●	●	●	●	●	●						●										
Investigate properties of multiplication/division															■																●	●
Explore number properties (commutative, zero, and identity)	●																															

Index